Engineering Maintenance Management

UNIVERSITY OF GLAMORGAN
LEARNING RESOURCES CENTRE

Pontypridd, Mid Glamorgan, CF37 1DL
Telephone: Pontypridd (01443) 480480

Books are to be returned on or before the last date below

INDUSTRIAL ENGINEERING

A Series of Reference Books and Textbooks

1. Optimization Algorithms for Networks and Graphs, *Edward Minieka*
2. Operations Research Support Methodology, *edited by Albert G. Holzman*
3. MOST Work Measurement Systems, *Kjell B. Zandin*
4. Optimization of Systems Reliability, *Frank A. Tillman, Ching-Lai Hwang, and Way Kuo*
5. Managing Work-In-Process Inventory, *Kenneth Kivenko*
6. Mathematical Programming for Operations Researchers and Computer Scientists, *edited by Albert G. Holzman*
7. Practical Quality Management in the Chemical Process Industry, *Morton E. Bader*
8. Quality Assurance in Research and Development, *George W. Roberts*
9. Computer-Aided Facilities Planning, *H. Lee Hales*
10. Quality Control, Reliability, and Engineering Design, *Balbir S. Dhillon*
11. Engineering Maintenance Management, *Benjamin W. Niebel*
12. Manufacturing Planning: Key to Improving Industrial Productivity, *Kelvin F. Cross*
13. Microcomputer-Aided Maintenance Management, *Kishan Bagadia*
14. Integrating Productivity and Quality Management, *Johnson Aimie Edosomwan*
15. Materials Handling, *Robert M. Eastman*
16. In-Process Quality Control for Manufacturing, *William E. Barkman*
17. MOST Work Measurement Systems: Second Edition, Revised and Expanded, *Kjell B. Zandin*
18. Engineering Maintenance Management: Second Edition, Revised and Expanded, *Benjamin W. Niebel*

Additional Volumes in Preparation

Engineering Maintenance Management

Second Edition, Revised and Expanded

Benjamin W. Niebel

Professor Emeritus of Industrial Engineering
The Pennsylvania State University
University Park, Pennsylvania

Marcel Dekker, Inc.   New York • Basel • Hong Kong

Library of Congress Cataloging-in-Publication Data

Niebel, Benjamin W.
  Engineering maintenance management / Benjamin W. Niebel. — 2nd ed., revised and expanded.
      p.   cm. — (Industrial engineering; 18)
  Includes bibliographical references and index.
  ISBN 0-8247-9247-5
  1. Plant maintenance—Management.   I. Title.   II. Series.
  TS192.N54   1994
  658.2'02—dc20   94–16961
                                                                                      CIP

The publisher offers discounts on this book when ordered in bulk quantities. For more information, write to Special Sales/Professional Marketing at the address below.

This book is printed on acid-free paper.

Marcel Dekker, Inc.
270 Madison Avenue, New York, New York 10016

Current printing (last digit):
10 9 8 7 6 5 4 3 2 1

PRINTED IN THE UNITED STATES OF AMERICA

Preface

The continuing effort by industries to be internationally competitive necessitates that industrial and business management become more cost-effective in the operation of all facets of an enterprise. Maintenance is an important area that has not been studied with the same intensity as, for example, production. Consequently, the maintenance effort in most companies can be improved significantly from the standpoints of quality, cost, and time of delivery.

Modernization has taken place in almost all industries and businesses. The installation of more flexible computer-controlled automatic equipment is commonplace. Also, in recent years, there has been an ever-increasing effort in the utilization of advanced mechanized material-handling equipment. With this modernization, the complexity and amount of maintenance have increased. It is not uncommon today to have the cost of maintenance exceed the cost of direct labor.

Because of this continuing increase in the importance of maintenance, the second edition of *Engineering Maintenance Management* has been written with the same objective as the first, that is, to provide practical assistance to the various levels of maintenance management. The first edition has been thoroughly tested by diverse industries throughout the world. It has been the basis of seminars and short courses offered not only in the United States but in Japan, South Korea, Singapore, Malaysia, Mexico, and China. The second edition should prove to be more beneficial to all those associated with maintenance—from the maintenance craft foreman to the chief plant engineer.

The second edition covers the fundamental theories, principles, procedures, and techniques utilized by engineering maintenance management. Throughout the book, proven techniques for providing more cost-effective maintenance are emphasized. It includes numerous illustrations, tables, charts, and forms valuable for establishing or improving procedures.

This edition contains new chapters on utilities management, plant rearrangement and minor construction, subcontracted maintenance services, fire protection, and controlling residual waste and contaminated storm water discharge. These materials have been added at the suggestion of the many users of this book. In addition, the chapter on designing for reliability has been expanded to include system effectiveness. This chapter—along with the chapters on preventive maintenance, diagnostic techniques, and computerized maintenance—provides the essentials for optimizing the maintenance function.

Beginning with the functional design of a new product and continuing to the identification of failure systems that permit planned maintenance, this book provides product and plant engineers with practical methods of developing more reliable products and systems that can be well maintained in a cost-effective manner. Application of the computer to provide more timely and thorough information for maintenance management control is emphasized. In addition, the "how" and "why" of maintenance management are covered thoroughly. There are up-to-date presentations on the maintenance organization, maintenance cost and control systems, estimating material and labor costs, inventory control of maintenance materials including economic order quantities, and maintenance planning and scheduling. Other chapters cover preventive and predictive maintenance and the utilization of diagnostic techniques. Substantive discussions are also presented in the areas of pre-employment testing of maintenance employees, training for maintenance work and engineering maintenance management work, measuring maintenance performance, determining compensation for maintenance work, and improving maintenance performance.

For their constructive criticism while I prepared this second edition, I wish to acknowledge Dr. E. Emory Enscore and Dr. Kenneth Knott of the Department of Industrial and Management Systems Engineering of The Pennsylvania State University. Acknowledgment also goes to D. F. Wanner, consultant manager of Maintenance Engineering, E. I. du Pont de Nemours &

Co., and D. L. Stairs, manager of Industrial Engineering, Floor Products Operations, Armstrong World Industries, for providing assistance with the first edition. I am also indebted to my wife, Doris M. Niebel, who typed the manuscript.

Benjamin W. Niebel

Contents

Preface   *iii*

1. Maintenance Organization   1

2. Maintenance Control Systems   14

3. Work Sampling in Maintenance   48

4. Estimating and Measuring Maintenance Work   64

5. Estimating Materials Costs in Maintenance Work   100

6. Inventory Control of Maintenance Materials   111

7. Maintenance Planning and Scheduling   125

8. Preventive and Predictive Maintenance   146

9. Diagnostic Techniques   189

10. Computerized Maintenance   205

11. Maintainability, Reliability, and System Effectiveness   230

12. Utilities Management   253

13. Plant Rearrangement, Minor Construction, and
    Subcontracted Services   275

14. Fire Protection and Controlling Residual Waste and
    Contaminated Storm Water Discharge   290

15. Training for Engineering Maintenance Management Work   299

16. Compensation for Maintenance Work   316

17. Measuring and Improving Maintenance Performance   339

Index   *363*

1
Maintenance Organization

The maintenance organization in any industry or business is faced with the same problems confronting manufacturing management in the production of a competitively priced product. The maintenance department can be thought of as a structured activity that is integrated with other departments of the enterprise and whose product is service.

Historically, the typical size of a plant maintenance group in a manufacturing organization ranged from 5–10% of the operating force (1 to 17 in 1969 and 1 to 12 in 1981). However, today the proportional size of the maintenance effort compared to the operating group has increased and is projected to continue to increase. This is because of the tendency for industry to increase the mechanization and automation of many processes, including the use of robots. This trend has decreased the need for operators and, at the same time, has resulted in a greater demand for technicians, electricians, and other service people. Note that we have said the proportional size of the maintenance effort will continue to increase but not necessarily the size of the maintenance organization. The writer, along with many other practicing engineers, has been recommending that production operators assume a larger share of the preventive maintenance effort (regular lubrication, oil and filter changing, cleaning of the facility, etc.) and the gathering of data (vibration signature, temperature, sound, color, etc.) for predictive maintenance. This is especially advantageous since the production worker is the closest individual to the equipment being monitored and is the individual who knows best when the behavior pattern of facilities is not normal. Thus, this author recommends the redesignation of production workers' duties to include a greater commitment to both quality and equipment maintenance.

Experience has proven that giving workers more flexibility and authority is the long-term solution to the vast majority of production problems. By adopting this philosophy, the size of the formal maintenance organization may well become smaller. In some cases, it may disappear altogether since it may be cost-effective to subcontract plant renovation, machine rebuilding, etc. when necessary and leave the day-to-day maintenance activities in the hands of production employees.

Maintenance is a dynamic activity comprised of a great number of variables interacting with one another, often in a random pattern. Industries and businesses that best manage the maintenance effort are cognizant of this dynamic randomness and develop structured maintenance systems to cope with it.

Since maintenance employees are high-priced people and the percentage of time spent performing useful maintenance work (i.e., excluding personal delays, walking, waiting, getting material, and getting information) is typically low (often less than 40%), it is becoming increasingly important in order for a business or industry to be competitive to have a sound engineered maintenance management program.

COST CONSIDERATIONS

Contrary to the opinion of many managers, maintenance costs are highly controllable. This is because

1. For the most part, maintenance is subjective. A great deal of maintenance or a minimal amount of maintenance can be performed without ceasing to operate.
2. Maintenance work is highly labor intensive and operations that are labor intensive can be controlled.

There are four fundamental costs associated with maintenance work. These are

1. *Direct costs.* Direct costs are those costs required to keep equipment operable. These include periodic inspection and preventive maintenance, servicing costs, repair costs, and overhaul costs.

2. *Standby costs.* The total cost of operating and maintaining standby equipment needed to be put in operation when primary facilities are either undergoing a maintenance activity or are unoperable for some reason.
3. *Lost production costs.* Costs due to lost production because primary equipment is down and no standby equipment is available.
4. *Degradation costs.* Those costs occurring in deterioration in the life span of equipment resulting from inadequate and/or inferior maintenance.

The reader should understand that more maintenance does not necessarily mean better maintenance. Too much maintenance can take needed production facilities out of operation to such an extent that cost-effective maintenance has been submerged. Figure 1.1 (upper left) illustrates the relationship between the quantity of maintenance and downtime cost.

Controlling direct maintenance costs will permit reasonable control of the four fundamental costs that are related to maintenance work. This is accomplished by thorough planning performed by competent maintenance technologists. This planning performs two functions prior to the actual making of maintenance repairs, preventive maintenance, and construction work, so that quality work can be done at the proper time and at the least cost. These two functions are method determination and screening. Method determination involves the selection of the best alternative as to how the repair or overhaul should be done. It includes the determination of the scope of the work requested and the preparation of a materials list required to perform the work. It also requires the preparation of a list of any special tools or facilities that may be needed to complete the work and any manpower specialists.

Screening is the process when the planner first decides, "Is the work really necessary?" and, if it is, he decides, "Where should it be performed?" The work may best be performed by a contractor or it may be cost-effective to have a portion of the work completed by a contractor and the remainder by the maintenance department. Or it may be best to perform the entire maintenance job internally.

THE RESPONSIBILITY OF MAINTENANCE

It is the principal responsibility of maintenance to provide a service to the operation that enhances its ability to make a profit. Thus the maintenance organization will have the objective of maximizing the availability of plant facilities in an operating condition permitting maximum performance and the

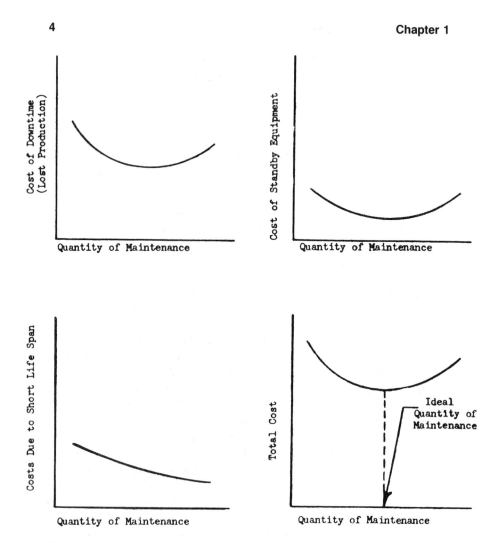

Figure 1.1 Relationship between the quantity of maintenance and various costs.

preservation of plant and equipment so as to provide a good service life. These objectives must be realized in a cost-effective manner.

There are corollary objectives that are related to those stated. These include providing a safe environment for all employees. A safe environment inevitably is a productive environment. Assistance in providing a clean environment is another responsibility. A clean environment enhances favorable working conditions which leads to better satisfied workers.

Figure 1.2 illustrates graphically the principal activities carried out by a well-organized engineered maintenance management system. Note that the integrated system endeavors to

1. Maximize the availability of production processes and equipment throughout the plant.
2. Conserve energy usage.
3. Conserve and control the use of spare parts and maintenance materials.
4. Perform all maintenance activities when needed, utilizing qualified employees that perform in an efficient manner.

Thus to manage a maintenance department effectively, one must be trained in the technology to be able to consistently know what work is required and

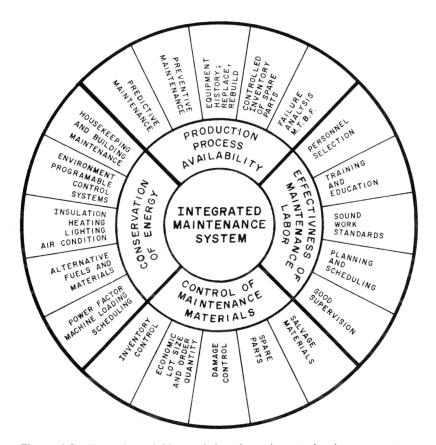

Figure 1.2 The major activities carried out by an integrated maintenance system.

to provide competent manpower with the right skills, tools, and equipment and the right parts and materials at the right time. The efforts of all involved workers must be so coordinated that a quality job is performed on schedule at the least possible cost.

THE PLACE OF MAINTENANCE IN THE PLANT ORGANIZATION

The place of maintenance in the plant organization will vary with the size, complexity, and product or products produced. Generally, the manager of maintenance will report to the plant engineer except in the small operation where the plant manager performs the function of plant engineering. Plant engineering in addition to maintenance should include the disciplines of reliability engineering, scheduling, standards development, and equipment inspection and evaluation. These disciplines function as staff organizations which facilitate the line maintenance organization. They will be discussed in detail later. Without this staff support, no maintenance program can be completely successful.

The plant engineer in turn will report to the top operating officer, who may have the title of vice president of operations or plant manager. On occasion, particularly in small and medium-sized plants, the plant engineer may report directly to the chief manufacturing officer, such as the vice president of manufacturing. Figure 1.3 illustrates the place of maintenance in a typical medium-sized to large plant. Those activities coming under the jurisdiction of the plant engineer permit effective engineered maintenance management. It is the coordination of these activities that allow the fulfillment of the objectives of modern engineered maintenance. It is these activities that we will be particularly concerned with in this text.

THE MAINTENANCE ORGANIZATION

The structure of the maintenance organization depends upon both the size and the product of the enterprise. Perhaps the first consideration in planning a maintenance organization is the determination of whether to have centralized or decentralized maintenance. If decentralized, then a decision needs to be made to determine if it is better to have the decentralization based upon area or department or a combination of area and department.

Certainly the majority of small and medium-sized enterprises that are housed in one structure, or service buildings located in an immediate geographic area, are best served by central maintenance. Also, central mainte-

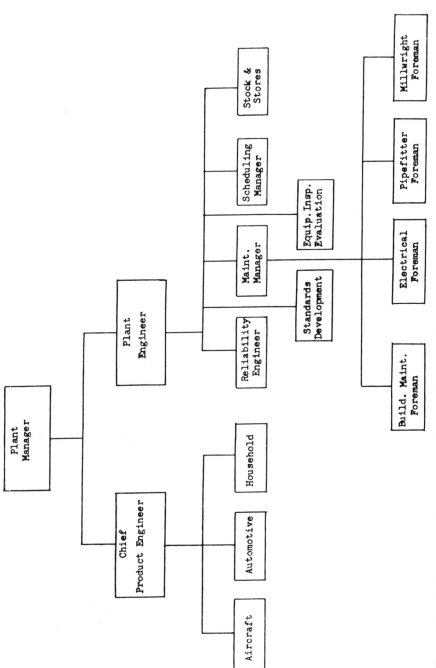

Figure 1.3 Medium-sized to large plant organization (12 to 20 craftsmen per supervisor).

7

nance usually is better when production workers assume the majority of the preventive maintenance, as advocated. The central maintenance department will be responsible for major repairs, equipment overhauling, building renovation, etc. Work assigned to the various maintenance craftpeople is channeled through the same head of maintenance. The obvious advantages of centralized maintenance include

1. Overall efficiency is higher than in decentralized maintenance, since greater flexibility prevails in assigning workmen with specific skills to assignments throughout the plant.
2. Fewer maintenance workmen are needed, since specially skilled craftsmen can be used in all maintenance areas.
3. Special equipment as well as specialized craftsmen are utilized to a higher degree.
4. Line supervision is usually more effective since one individual is responsible for all maintenance. This individual, since we are assigning but one, is usually more highly selected and trained than his counterparts in a decentralized organization.
5. Centralized maintenance usually permits more effective on-the-job training.
6. Centralized maintenance permits the procuring of more modern facilities.

The principal disadvantages of centralized maintenance are

1. More time is taken getting to and from the job.
2. Supervision of work is more difficult in view of the remoteness of the maintenance in relation to the centralized headquarters.
3. Since different specialists can be assigned to the same capital equipment (over time), no one individual becomes completely familiar with complex hardware. Thus the slope of the manufacturing progress function for complex equipment is less than under decentralized maintenance. (Manufacturing progress function is the progress in production effectiveness with the passing of time.)
4. More cost in connection with transportation is necessitated in view of remoteness of some of the maintenance work.

Under decentralized maintenance, a separate maintenance group is assigned to either a specific area or some unit such as a department. The major reasons for decentralized maintenance are the reduced travel time in getting to and from the job and the spirit of cooperation that exists between production and maintenance workers when working together in the same generalized area.

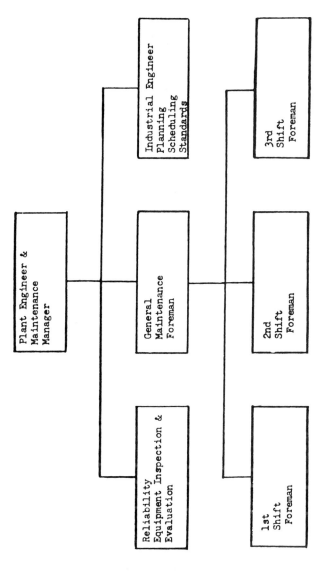

Figure 1.4 Maintenance organization chart utilizing centralized maintenance typical of small and medium-sized plants.

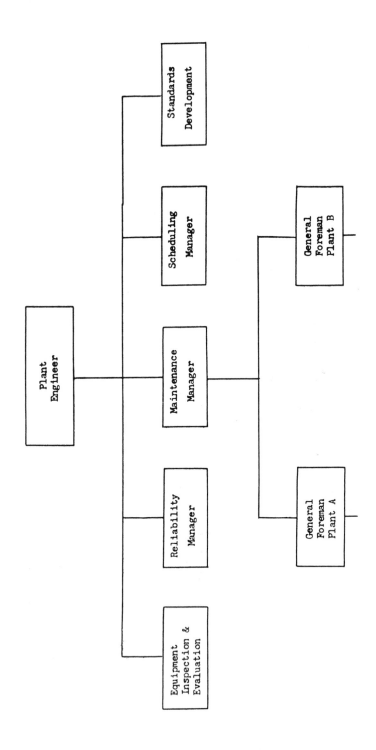

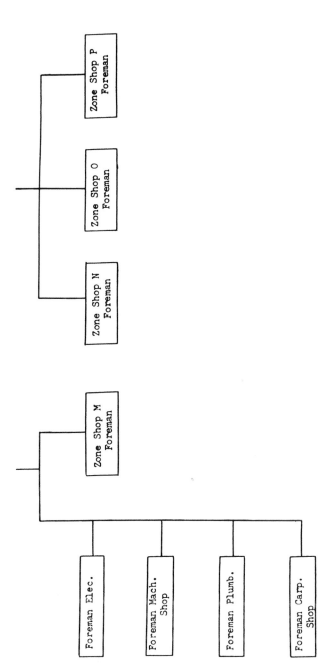

Figure 1.5 Maintenance organization chart combining centralized and decentralized maintenance. This type of organization is typical of large organizations.

11

Also, supervision is usually closer and workmen become more familiar with complex sophisticated facilities in view of the same specialists being reassigned to service the same equipment. Decentralized maintenance loses its principal advantage when production workers assume a greater responsibility for preventive maintenance. For this reason, it is only in large plants that decentralized maintenance should be used—where it is cost-effective to have major overhauls, facility rebuilding, etc., take place in different locations so as to reduce travel time.

In large and very large plants a combination of both centralized and decentralized maintenance will usually work the best. In this way the advantages of both systems can be achieved with very few of the disadvantages. Figures 1.4 and 1.5 are representative organization charts showing maintenance organizations in typical small, medium-sized, and large plants. For example, some areas of the organization may justify a zone maintenance shop if the importance of the unit justifies assigned personnel to be available immediately as equipment develops symptoms of failure or if there is a sufficient workload to keep certain specialized mechanics working the most productive time as opposed to nonproductive time each day.

From the preceding discussion, it should be apparent that no one type of maintenance organization is best suited to all types of plants. It is important that every enterprise establish an organization that effectively provides lines of communication through which the responsibilities of maintenance are carried out. These responsibilities include the planning, scheduling, installing, supervision, performance, and quality of maintenance in all plants and equipment facilities. The organization will clearly identify who administers and coordinates manpower, materials, tools, and supplies related to maintenance and preventive maintenance.

SUMMARY

In order to fulfill the objectives of the maintenance function as outlined in this chapter, the enterprise needs (1) management skills to integrate people, policies, equipment, practices, and to evaluate the maintenance performance and (2) engineering and technological skills in order to provide the best possible preventive maintenance, repair, and overhaul of the ever-increasing automatically controlled production equipment characteristic of modern day technology.

Today, the maintenance activity is typically headed by a manager who integrates the plant engineering function. Consequently, line maintenance is facilitated by various staff, including technologists and engineers, in work related to planning, job scheduling, cost estimating, standards development, reliability, and equipment inspection and evaluation. The line maintenance organization is usually supported by a staff of supervisors organized in larger establishments on a craft basis, and in smaller ones on a departmental or area basis. These supervisors then manage a team of hourly employees of the same craft and of different grades in larger organizations or of different crafts and different grades in smaller plants.

In planning the maintenance activity, management must decide to whom it should report and whether centralized or decentralized maintenance or a combination will be most effective for the operation. Historically, maintenance has reported either to the plant manager (sometimes referred to as the manager of operations) or the production manager, or to plant engineering.

In view of the present trend of movement to automatically controlled production, where equipment is operated at higher speeds and products are produced with closer tolerances, the maintenance function is growing not only in numbers of people employed, but in quality of employees. It is therefore strongly recommended that the maintenance function report either to plant engineering or the plant manager so that the activity can take full advantage of the technological support that is so important to the maintenance function.

Centralized maintenance will provide better control and more efficient staffing while decentralized maintenance has the advantage of specialization and usually faster turnaround time of the repair and overhaul functions.

In small and medium-sized companies, centralized maintenance is recommended. In larger and more complex companies, a hybrid of centralized and decentralized maintenance is recommended with decentralization taking place only when either (a) the work load is such to keep specialists working optimum time, (b) the nature of the facility or facilities is such that immediate maintenance or repair is needed in order to keep the equipment running, or (c) it is the most cost-effective method of organization.

SELECTED BIBLIOGRAPHY

Heyel, Carl. *The Encyclopedia of Management*, 3rd ed. New York: Van Nostrand Reinhold, 1982.
Mann, Lawrence, Jr. *Maintenance Management*. Lexington, MA: Heath, 1976.
Nakajima, Seiichi, Yamashina, Hatime, Kumagai, Chitoku, and Toyota, Toshio. "Maintenance Management and Control." In Gavriel Salvendy, Ed., *Handbook of Industrial Engineering*, 2nd ed. New York: Wiley, 1992.
Newbrough, E. T. and the staff of Albert Ramond and Associates, Inc. *Effective Maintenance Management*. New York: McGraw-Hill, 1967.

2
Maintenance Control Systems

The heart of sound engineering maintenance management is the control system. This control system must clearly identify what work is to be done, what materials are needed, when the work should be done, how long it should take, what skills are needed to perform the work, and what special tools are needed. The system should permit regular reporting of accurate records so that projected completion schedules are maintained and quality of the maintenance work is assured. Finally, the system should capitalize on the work accomplished by making improvements. Figure 2.1 illustrates schematically what is embraced in a maintenance control system. This is illustrated by those programs, records, reports, and evaluations that are contained within the dotted square. The various programs that facilitate plant availability include predictive and preventive maintenance, emergency maintenance, reliability improvement, cost reduction, and training. Records and reports that need to occur as production takes place include maintenance performance, product quality, equipment failure, equipment history, and costs. These data permit an analysis and evaluation so that the various engineering maintenance management programs can be improved, thus allowing greater plant availability.

The reader should understand the distinction between *predictive* and *preventive* maintenance.

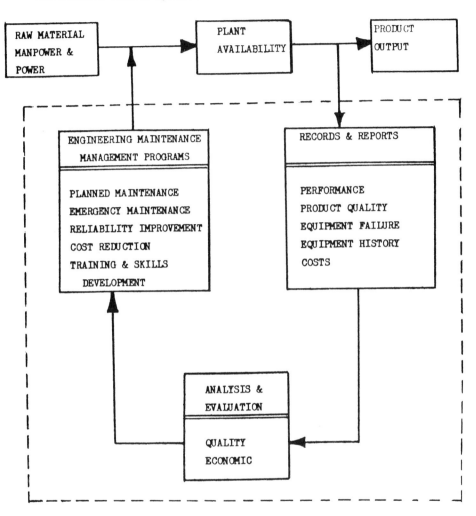

Figure 2.1 Activities included in a sound maintenance control system.

Predictive maintenance is that maintenance that takes place in advance of the time a failure would occur if the maintenance were not performed. The time when this maintenance is scheduled is based upon data that can be used to predict approximately when failure will take place if certain maintenance is not undertaken. Data such as vibration, temperature, sound, color, and others, provide information as to the best time to perform the predictive maintenance, which can involve repair, replacement of parts, and overhaul.

Preventive maintenance (PM) is that maintenance that is performed regularly to assure the cost-effective operation of the equipment. Preventive maintenance includes regular lubrication, replacement of filters, belts, fluids, ignition parts, valves, among others.

MAINTENANCE WORK ORDER SYSTEM

Although the terms ''work request'' and ''work order'' are often used synonymously, we will use the term work order to identify a request that has been screened by the planner who has decided that the work is necessary and who will perform the work. A number is assigned and the work is scheduled.

A work request is a written communication from a supervisor to have some work done. A written request form should be completed for all jobs.

The purpose of the maintenance work order system (this includes both the work request and the work order) is to provide a method for

1. Requesting in writing that work be performed by the maintenance department
2. Assigning the best method and craftsmen to perform the work in an estimated amount of time and cost
3. Reducing costs through man-hours and materials control of all maintenance work
4. Performing predictive and preventive maintenance
5. Improving the planning and scheduling of maintenance work

Responsibility for administration of the maintenance work order system should lie with the individual who is in charge of the planning and scheduling of maintenance work. The forms designed should emphasize clearly and with brevity what is to be done and how it is to be done. Figure 2.2 illustrates a combination work request/order form. Typically, this form would be completed in quadruplicate, where the first copy (white) is sent to the accounting department, the second copy (yellow) is the job assignment ticket, the third copy (blue) is sent to the requesting department, and the fourth copy (green) is filed in the maintenance department. It is helpful to include a part failure code on the work order form. This code will identify both the subassembly and part that needs repair as well as the reason for the repair and the action taken. In order to identify the subassembly and particular part a representative code card can be developed similar to that illustrated in Fig. 2.3.

DATE REQUESTED:	REQUESTED BY:	FROM DEPT.	JOB REQUIRED	AUTH. BY	CHARGE TO:

			TODAY	OTHER DATE		
			THIS WK			
			NEXT WK			

WORK REQUESTED

EQUIP. #

PROJECT #

WORK TO BE PERFORMED

EMERGENCY ☐

SCHEDULED ☐

DATE COMPLETED

ACCEPTED BY

DATE CLOSED

MATERIALS & SPECIAL EQUIP.

CLASSIFICATION ☐☐☐☐

REGISTER
#

FAILURE CODE						
SUB-ASS'Y	PART	CAUSE	ACTION	EST. HRS.	ACTUAL HRS	WORKER

PREVENTIVE
REPAIR
OVERHAUL
SAFETY
NEW
CONSTRUCTION ☐

Figure 2.2 Work request/order form.

Action Taken

Repair ☐   Replace ☐   Overhaul Order ☐

Reason for Failure

01 contamination
02 fatigue
03 heat
04 misalignment
05 normal wear
06 inadequate lubrication
07 setup improper
08 overload
09 short circuit
10 out of balance
11 other

Sub Assembly

01 apron
02 arm assembly
03 air control
04 battery
05 base
06 bridge
07 brake
08 breaker
09 bed
10 clutch
11 controller
12 coolant system
13 cross slide
14 crank shaft
15 cylinder
16 distributor
17 drive
18 drum
19 fan
20 face plate
21 feed mechanism
22 furnace
23 housing
24 hoist
25 knee
26 motor
27 pump
28 push button
29 ram
30 record and control
31 saddle
32 spindle
33 stock feed
34 switch
35 timer
36 tap changer
37 turret
38 valve
39 other

Specific Part

01 adjusting screw
02 arm
23 coupling
24 connecting rod
45 hopper
46 hose
67 shaft
68 sheaves

03 armature
04 band
05 beam
06 bearing
07 bells
08 belt
09 blade
10 bracket
11 brushes
12 bushing
13 cable
14 cam
15 cam shaft
16 chain
17 clamp
18 clapper
19 collector
20 commutator
21 connecting rod
22 coil

25 cups
26 crank shaft
27 diaphragm
28 disc
29 dog
30 drive shaft
31 fan
32 filter
33 finger
34 fork
35 fuse
36 gasket
37 gear
38 gib
39 guide
40 guard
41 gage
42 handle
43 heater element
44 hock

47 horn
48 impeller
49 lever
50 lamp
51 lead
52 nut
53 nozzle
54 packing
55 plates
56 plunger
57 plug
58 piston
59 pulley
60 rack
61 ram
62 resistor
63 retainer
64 rings
65 rolls
66 rotor

69 screw
70 slide
71 solenoid
72 socket
73 sleeve
74 spring
75 strainer
76 terminals
77 thermocouple
78 thermostat
79 tires
80 toggle
81 tongs
82 tubing
83 valve
84 ways
85 wire
86 wheel
87 worm
88 other

Figure 2.3 Code card for machine tools.

19

It should be understood that preventive maintenance usually would not be carried out by this request form (even though space has been assigned to check in the event some preventive maintenance work was required). Preventive maintenance operations, such as cleaning, lubrication, and regular replacement of components, would be handled by the issuance of standing work orders. These standing orders are issued periodically (usually monthly) and provide scheduling of where, when, and what is to be done to specific facilities and equipment.

The form illustrated (Fig. 2.2) is applicable for the majority of repair work. However, for major overhaul and new construction work a more detailed work order form giving instruction as to the crafts required to complete the job along with estimated standard hours and materials used should be employed. Figure 2.4 illustrates such a work order form.

The reader should understand that maintenance work order systems need to be tailored for the specific business or industry under consideration. Forms developed in a foundry would have many deficiencies if applied in a rubber fabrication plant. Some general guidelines that apply to all systems follow.

1. Work order forms should be numbered and at least three copies are needed in connection with the control system.
2. Work requests can be initiated only by authorized supervisors approved by management. A list of names of those designated should be furnished to the work order section of the maintenance department.
3. Work orders are initiated only in the maintenance department.
4. The work request and/or work order form should be the basis to transmit information to all crafts concerned (this may require extra copies).
5. The work order is the basis for reporting time and material charges (the work order number being used for this purpose).
6. The work order is the basis for accumulating records of job estimates and actual costs for control and for improving estimates.
7. The work order is the basis for work backlog reporting to control the size of the maintenance work force.

WORK ORDER DISPATCHING PROCEDURE

Although there is no single dispatching procedure that is applicable to all situations, it is fundamental that an orderly system be incorporated so that all interested parties are informed of the work to be done in adequate time and records can be maintained for both cost control and cost reduction.

WORK ORDER NO. 7-9954	EQUIPMENT SAND MIXER #2 LINE	ASSET NO. 5736
DATE 11/1	PART/SECTION MOTOR 75 HP	NO.
NAME C.TABOJKA	PROBLEM BURNT OUT	
AUTHORIZATION ① 2 3 4 5 S		
AVAILABLE / / TIME	SUGGESTION REPLACE MOTOR	
REQUIRED / / TIME		
PREMIUM TIME ☑ YES ☐ NO		

MATERIAL AND TOOLS REQUIRED

QUANT.	DESCRIPTION	PRODUCT SYMBOL	PRICE UNIT	TOTAL	LOCATION	ORD. NO.
1	MOTOR 75 HP				HOT. STORAGE	

NO.	TRADE	MEN	WORK DESCRIPTION	COMP.
1	EL	1	DISCONNECT	.4
2	MW	2	REMOVE AND REPLACE BELTS	.7
3	MW	2	REMOVE AND REPLACE MOTOR	2.0
4	EL	1	CONNECT	.4

TRADE	MEN	HOURS PLAN.	EST.	PLANNER
EL	1	1.2		T.T.
MW	2	4.0		J.L.

DRAWING NO.

ⓂM IM C/A AFE PM S
ARR.   SCHED.
MACH. REC. ☑ YES ☐ NO
MAJOR 556   SUB 40

Figure 2.4 Sample maintenance work order.

The maintenance work order dispatching duties and responsibilities include the following:

1. Upon receipt of work order request (this can be initiated by computer terminal, telephone, personal visit, or in written form), a work order is completed for each favorably screened request and includes the following information:
   a. Requestor's name
   b. Department number where work is to be performed
   c. Machine or equipment numbers
   d. Location of machine or equipment
   e. Maintenance department or department that will be doing the work
   f. Complete description of repair work required
   g. Date work is to be completed
   h. Date request received (including time of day)
   i. Date work order issued
   j. Crafts employed and standard hours to perform the work
2. Assign a priority to each work order. Usually three levels are considered: emergency, rush, and scheduled.
3. Maintain a work order register that lists pertinent data for each work order processed. Figure 2.5 illustrates a typical work order register.
4. File by order number one copy of the work order in the maintenance control department. An additional two copies are sent to each section or department of the maintenance activity that is involved.
5. Develop a weekly tabulated report showing the hours worked by jobs and by maintenance departments.
6. When one of the two copies of the order is returned (item 4 above) by the maintenance foreman indicating that the work has been completed, the hours are summarized for each completed job and the total hours are posted on both the return copy and the copy filed in the maintenance department by order number. These two copies of the completed order will be filed in the maintenance department: one copy will be filed by machine or equipment number and the other by date of completion.

MAINTENANCE DEPARTMENT'S PROCEDURE

As discussed earlier, two copies of each maintenance work order (these are usually different colors) are routed to the maintenance department foreman. One copy is for the use of the maintenance foreman and the other copy is given to the person assigned to the work.

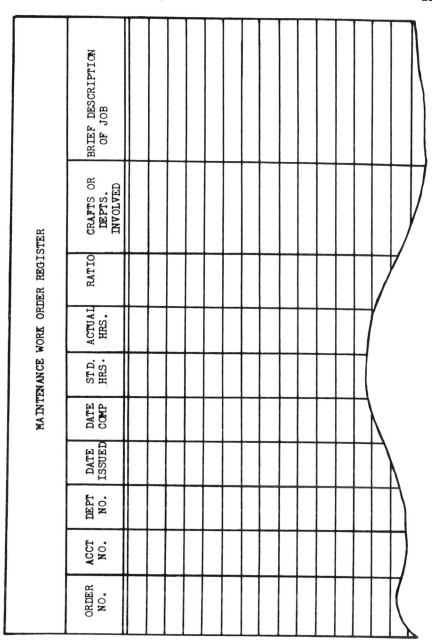

Figure 2.5 Maintenance work order register.

The maintenance foreman, typically, will maintain three open files. These are

1. Orders not started
2. Orders in process
3. Orders completed

When a job is scheduled, the maintenance foreman will place the workman's copy of the order on a clipboard or in a pigeon hole file segregated according to maintenance employee. This copy of the order serves as instructions to maintenance employees. If it is found that additional work is required, this information should be noted on this copy of the work order.

If it is found that an additional maintenance department is required to complete the work, the maintenance foreman shall notify the work order dispatcher who will issue the necessary additional copies to the department involved. Copies of the work order form should be available with no number preprinted, which will be used for this purpose and the original order number will be written in the space provided.

TIME REPORTING

In order to control costs and provide information leading to method improvement, it is essential that an accurate reporting of time be included in the maintenance control system. Figure 2.6 illustrates a job card that can be issued

ORDER #	EST. HRS.	ACCOUNT #	PROPERTY #	DEPARTMENT		DISPATCHER	NATURE OF WORK	DATE	ORDER #
EMPLOYEE NAME		WORK PERFORMED							
MASTER #									ASSIGNED TO
FUNCTION									
APPROVAL									
SHIFT A B C	ACT. HRS.								
		OUT IN INSTRUCTIONS							
☐ JOB COMPLETE	☐ TRADE COMPLETE	☐ TRADE INCOMPLETE	☐ NO STOCK	☐ CONVERT TO WORK ORDER					

Figure 2.6   Job card.

to each employee participating on each job. The form shown is compatible for both manual posting and electronic data processing. The pertinent information included on the job card is order number, date, employee name, employee clock or master number, department, shift, actual hours utilized, time work begun, time work completed, order status, description of work, and approval of the line supervisor.

The work order can be used to record the time spent on the job instead of the job card as mentioned previously. This information can be supplied by the worker or the foreman. This method of time reporting is effective only when each worker who works on the job is given a copy of the work order.

In some instances each workman maintains a daily work timecard where the time spent on each work order is recorded. See Fig. 2.7.

THE INFORMATION FLOW

The flow of paperwork in conjunction with the work request/work order system will vary depending upon the size of the maintenance organization, the type of organization (centralized or decentralized), the type of industry, and the effectiveness of the management team. Figure 2.8 illustrates the flow of activity in a typical medium-sized plant.

Included in the flow of information are the computer systems, of which there are five normally used in the total maintenance control system. These are equipment control, work control, inventory control, cost reporting, and management reporting. These systems control the important parameters of equipment, labor, and materials. They calculate the cost and measure the performance and make several analyses in order to report to management the effectiveness of the integrated maintenance program.

MAINTENANCE REPORTS

There are several records and reports to management that should be completed regularly. Notable among these are performance, product quality, equipment failure, equipment history, and costs. This information and the resulting actions initiated are fundamental to any maintenance control system. With sound engineering maintenance management, two objectives will be realized. There will be improvement in the plant availability profile and there will be improvement from the standpoint of maintenance costs per unit of plant output. Thus management should expect and should get both better maintenance and less expensive maintenance as the engineering maintenance management system matures.

TIME RECORD

EMPLOYEE NAME---CLOCK NO.-----------

WEEK ENDING--SHIFT----------------------

FOREMAN APPROVAL----------------

ORDER NO.	M	T	W	TH	F	S	SU	TOTAL
TOTAL								

Figure 2.7 Daily work time card.

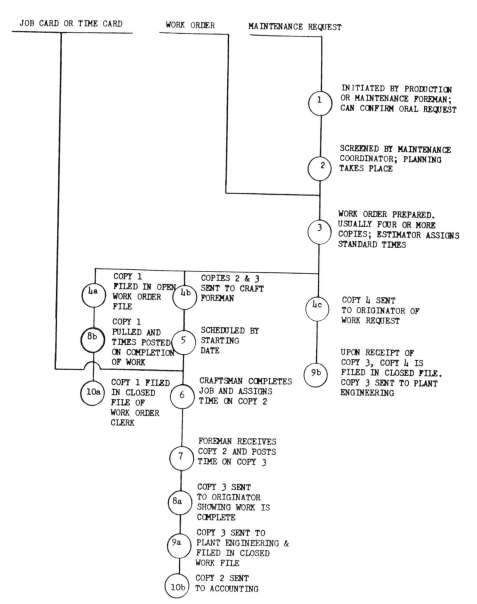

Figure 2.8 Flow of maintenance activity in a typical medium-sized plant.

THE MANUFACTURING PROGRESS FUNCTION

Since learning is time-dependent, we can expect the planning function of maintenance work to become more and more efficient as well as the craftsmen who do the actual work. Furthermore, methods engineering as undertaken by industrial engineering will continue to make improvements in both how maintenance work shall be accomplished and in the materials utilized.

The theory of the manufacturing progress function, sometimes referred to as the learning curve, proposes that as the total quantity of similar or identical work doubles, the time per unit declines at some constant percentage. For example, if it is expected that an 85% rate of improvement will be experienced, then as production doubles the cumulative average time per unit will decline 15%.

Table 2.1 illustrates the decline in the cumulative average hours per unit of production with successive doubling of the maintenance quantity where an 85% rate of improvement exists.

When linear graph paper is used, the manufacturing progress function is a hyperbola of the form: $y_x = kx^n$. On log-log paper, the curve is represented by

$$\log y_x = \log k + n \log x$$

where

y_x = cumulative average value of x units of maintenance work

k = time to perform the first unit of maintenance work

x = number of units produced

n = exponent representing the slope (tangent ϕ in Fig. 2.9)

Table 2.1 Decline in Cumulative Average Time Under an 85% Manufacturing Progress Function

Cumulative maintenance of same or similar work	Cumulative average hours per unit of maintenance work	Ratio to previous cumulative average
1	100	—
2	85	85
4	72.25	85
8	61.41	85
16	52.20	85

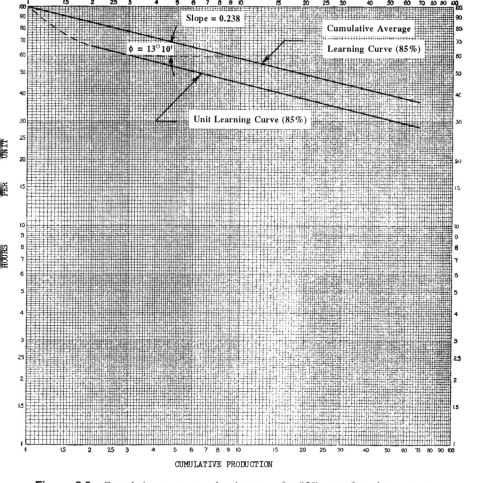

Figure 2.9 Cumulative average and unit curves for 85% manufacturing progress function.

By definition, the manufacturing progress function in percent is then equal to

$$\frac{k(2x)^n}{k(x)^n} = 2^n$$

Taking the log of both sides, n = log of manufacturing progress/log 2. For 85% manufacturing progress, n = log of 0.85/log of 2 = -0.2345. In Table 2.2, the slopes of common manufacturing progress percentages are provided.

When working with the man-hours required to perform a specific repair or overhaul, we are dealing with the *unit manufacturing progress function*, which refers to the hours required to repair a specific unit. The log plot of the cumulative average is asymptotically parallel to the log plot of the unit curve. The cumulative average line is straight, while the individual line curves downward from unit one until it becomes parallel to the cumulative average line.

To plot the unit time versus the quantity, two points may be calculated and the plotting made on log-log paper. To calculate the unit time value of the selected points, multiply the cumulative average time of these points by a conversion factor. The conversion factor used for making the unit plot is $1 + n$. This is obtained as follows:

$$y_x = kx^n = \text{cumulative mean for } x \text{ planes}$$

$$T = xy_x = kx^{n+1} = \text{total time for } x \text{ planes}$$

Since the time for each individual plane is a function of x, $f(x)$ = time for each plane.

$$T = xy_x = \int_0^x f(x)\, dx = kx^{n+1}$$

Differentiating $dT/dx = f(x) = (n + 1)kx^n = (n + 1)y_x$ = time for xth plane.

Table 2.2 Slope Values for Representative Manufacturing Progress Functions

Manufacturing progress function percentage	Slope
70	0.514
75	0.415
80	0.322
85	0.234
90	0.152
95	0.074

Thus the conversion factor for an 85% learning curve would be

$$1 + n = 1 + (-0.234) = 0.766$$

In this way it is possible to estimate the time it should take for each repair of like facilities under a multiple repair or overhaul situation once we measure or estimate the time for the first repair and know what manufacturing progress function prevails. Figure 2.9 illustrates the plotting of the cumulative average overhaul time and the unit overhaul time for an 85% manufacturing progress function and the time of 100 man-hours to overhaul the first machine.

PERFORMANCE REPORTS

There are three classes of reports dealing with performance that will prove to be useful in helping to assure the success of the maintenance control system. These are (1) maintenance costs per some unit of production output, (2) plant availability, (3) maintenance efficiency reports.

The first class of report provides a maintenance cost profile of the various departments or cost centers. Figure 2.10 illustrates such a report. Note that for the month of March, the machine shop (cost center 410) consumed a total maintenance cost of 45% of the value of the direct labor output. This 45% figure will, of course, be compared to corresponding values for previous months. This figure should decrease progressively with the maturing of the maintenance control system.

The second class of report provides a plant availability profile by department or cost center. For example, Figure 2.11 illustrates such a report. Note that for the month of March, 13% of the capacity of the machine shop was utilized for scheduled and unscheduled maintenance and breakdowns.

The third type of report assumes that the maintenance control system includes standards that are assigned in advance of actual repair, service, and overhaul. Table 2.3 illustrates a typical report that shows performance by maintenance craft. In this report, it is a good idea to show what proportion of the total hours worked by department were worked on an overtime basis. Needless to say, it is good management practice to have control over the amount of overtime. Other helpful reports in this category include maintenance backlog; distribution of maintenance hours performed as to percent on rush jobs, percent on PM inspection, and percent on minor servicing; and percent of repair jobs originated as a result of PM inspection.

In the maintenance efficiency category, it is good practice to maintain a backlog report by craft. The engineered maintenance control system can only be effective if the number of employees within each craft has a direct relation

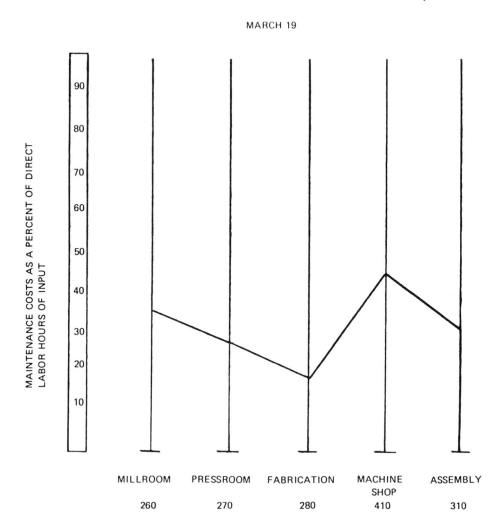

Figure 2.10   Maintenance cost profile.

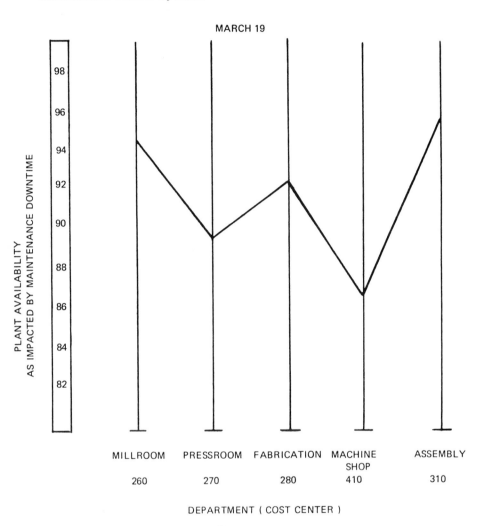

Figure 2.11 Plant availability profile.

to the backlog of requested work. Too little backlog will result in performance below standard, craft waiting time, and inefficient scheduling. Too large a backlog may result in inferior maintenance practice and poor service to the production departments.

A backlog of work ranging between two and four weeks at normal performance is generally considered to be satisfactory. With a work scheduling

Table 2.3 Maintenance Craftsmen Performance Report

Craftsmen	Regular hours	Overtime hours	March 19				Remarks
			Total hours worked	Standard hours	Percent variance		
Carpenters	692	10	702	755	+7.5		
Millwrights	1384	—	1384	1540	+11.3		
Plumbers	1038	50	1088	1050	−3.5		Breakdown in steamroom caused significant overtime
Electricians	865	—	865	900	+4.0		
Machinists	3460	190	3650	4020	+10.0		
Sheet-metal workers	346	—	346	400	+15.6		Breakdown in steamroom caused significant overtime
General laborers	1730	60	1790	1850	+3.3		

system in operation (see Chapter 7), it is a simple matter to generate periodically (usually every week) a maintenance work backlog report. When the backlog report indicates a downward trend, action should be taken. This includes one or more of the following:

1. Reduce outside contracted work.
2. Transfer between crafts.
3. Transfer between departments.
4. Reduce the maintenance force.

When the maintenance backlog keeps rising, again action should be taken. In this situation, the same sequence is followed.

1. Increase the outside contracted work.
2. Transfer between crafts.
3. Transfer between departments.
4. Schedule cost effective overtime.
5. Increase the maintenance force.

A maintenance efficiency report that is quite helpful is one that regularly (once a month is the usual period) shows the percent of time spent on rush or emergency (nonscheduled) jobs and the percent of time spent on repair jobs originated as a result of PM inspections, and finally the percent of man-hours on PM inspection or minor servicing jobs. This information can be included on the craft performance report (see Table 2.3). All of this information can be obtained regularly from the work order data processing system.

The percentage hours on rush jobs (and particularly overtime hours) should diminish as the maintenance control system develops. The information on the amount of repair resulting from PM and the amount of time spent on PM will permit the cost justification of the preventive maintenance system.

PRODUCT QUALITY REPORTS

Although we usually think of product quality in relation to direct labor and quality control, it can be understood easily that quality of output is as much a function of the quality of the processing facilities as it is to the skill and the experience of the operator. For example, the capability of a machine tool that is in first-class condition may have a tolerance distribution similar to that illustrated in Fig. 2.12. Here, 99% of the time, the machine will produce parts within ± 0.0005 in if the operator has centered his tooling properly. However, after the machine has been in service for some time and wear has taken place in the bearings and other moving parts, there will be considerably more chatter

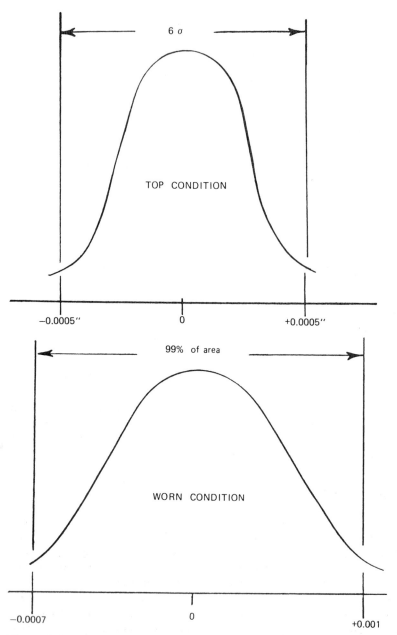

Figure 2.12   Distribution of tolerance capability on a new and a worn facility.

and vibration. The tolerance of this same facility with the tooling properly centered may degenerate to the distribution illustrated in Fig. 2.12 (bottom). A monthly report giving information as to the number of rejects or the product quality of the major operating equipment will help identify when an investigation or precision inspection is needed to determine the cause and take appropriate corrective action, which may be a repair or major overhaul of capital equipment.

EQUIPMENT FAILURE REPORTS

Just as it is important to have regular reporting on product quality, it is equally important to be knowledgeable as to the fundamental characteristics of the equipment failures being experienced. Figure 2.13 illustrates a failure report. The information that should be recorded on this report after a failure has occurred should be carefully analyzed so that corrective action can be taken to minimize the chance of recurrence of such failure. This analysis is usually performed by the Chief Plant Engineer in concert with the Maintenance Supervisor and the Reliability Engineer who participated in the equipment failure report.

EQUIPMENT HISTORY

A fourth record that needs to be kept in order to maintain a sound control system is a history of the production facilities. This record includes a history of the repairs performed as well as cost and plant engineering specifications. This information is useful in determining when equipment should be replaced and the nature of the replacement equipment. Figure 2.14 illustrates a typical equipment history record. On the front, physical plant information is posted, while the back shows a record of the maintenance performed.

COSTS

In order to check the performance of each work order, a cost clerk will compute the total cost required to complete the job. You will recall from the flow of paperwork a copy of the work order is sent to accounting (see Fig. 2.8). Here all labor at actual rates (regular and overtime) are extended and totalized to give the cost of labor. On materials, unit prices from purchasing (if not already entered on the work order) are obtained. A fixed overhead rate, determined by accounting, is applied percentagewise to the labor cost. The total cost of the maintenance performed is the sum of all three.

EQUIPMENT FAILURE REPORT	
EQUIPMENT NAME	NUMBER
SHUTDOWN TIME & DATE PROBABLE CAUSE	
EFFECT ON OTHER EQUIPMENT & OPERATIONS	
STANDBY EQUIPMENT NAME & NUMBER	
INITIATED BY	DATE

THIS SECTION TO BE COMPLETED BY MAINTENANCE SUPERVISOR

REASON FOR FAILURE

ORDER #	DATE STARTED	DATE COMPLETED	SUB-CONTRACTED		TOTAL HOURS
			ON SITE		
			WORKSHOP		

ACTION TAKEN

RECOMMENDATION TO PREVENT SIMILAR FAILURE

SIGNED   DATE

THIS SECTION TO BE COMPLETED BY RELIABILITY ENGINEER

PROBABLE REASON FOR FAILURE

ACTION TAKEN

RECOMMENDATION TO PREVENT SIMILAR FAILURE

SIGNED   DATE

Figure 2.13 Typical equipment failure report.

A summary of maintenance costs by work order is issued monthly. This information is used not only to assist in the control of future maintenance costs by the department, but also to develop indirect cost ratios that will accurately predict the cost of the manufactured product.

```
┌─────────────────────────────────────────────────────────────────────┐
│                         EQUIPMENT  HISTORY                            │
├─────────────────────────────────────────────────────────────────────┤
│ DESCRIPTION                                                           │
├─────────────────────────────────┬──────────────┬────────────────────┤
│ MFG                             │ MODEL        │ TYPE               │
├──────────────┬──────────────────┼──────┬───────┴────────────────────┤
│ CAPACITY     │ WEIGHT           │ FLOOR AREA                         │
├──────────────┴──┬───────────────┴────────────────────────────────────┤
│ SERIAL #        │   AIR  ☐        STEAM    ☐                          │
│                 │   GAS  ☐        SEWER    ☐                          │
│                 │   WATER ☐       ELECTRIC ☐                          │
├─────────────────┴──┬──────────┬───────┬──────────────────────────────┤
│ MOTOR MFG.         │ MODEL    │ H.P.  │ SERIAL #                     │
├──────────┬─────────┼──────┬───┴───┬───┴──────┬─────────────────────── │
│ FRAME    │ R.P.M.  │ VOLTS│ PHASE │ CYCLE    │                        │
├──────────┴──┬──────┴──────┴───┬───┴──────────┴────────────────────────┤
│ PURCHASE PRICE │ FREIGHT      │ INSTALLATION COST                     │
├──────────────┬─┴──────────────┼───────────────────────────────────────┤
│ DATE ACQUIRED│ INVENTORY #    │ LOCATION                              │
├──────────────┴────┬───────────┴───────────────────────────────────────┤
│ WARRANTY          │ MTBF                                              │
├───────────────────┴───────────────────────────────────────────────────┤
│                      SPARE  PARTS  STOCKED                            │
├──────────────────────────────────────────┬─────────┬─────────┬────────┤
│  PARTS  DESCRIPTION  &  SUPPLIER          │ PART NO.│ QUANTITY│UNIT COST│
│                                           │         │         │        │
│                                           │         │         │        │
└──────────────────────────────────────────┴─────────┴─────────┴────────┘
```

Figure 2.14a   Equipment history report (front).

ENGINEERED MAINTENANCE MANAGEMENT PROGRAMS

There are five engineered maintenance management programs whose combined objective is to improve the plant availability at the least cost and allow production of the highest quality (see Fig. 2.1).

Planned Maintenance

The more effective the maintenance control system, the greater the percentage of total maintenance is planned. It has been estimated that over 90% of plant

DATE	MAINTENANCE PERFORMED	MATERIAL	LABOR	LOST PRO. HRS

Figure 2.14b   Equipment history report (back).

maintenance and construction can be handled on a planned basis. Planning defines what is to be done and how it is to be done. It specifies the materials, tools, equipment, and skills required to perform the work. In small to medium-sized plants, the planning function is often handled by the maintenance supervisor. An obvious advantage here is that the supervisor is much better prepared to manage the job properly. Where someone other than the supervisor handles the planning function, it is important that he or she has a practical maintenance background equivalent to that of a maintenance supervisor. A technical background equivalent to a B.S. degree in engineering is also highly recommended.

Planned maintenance represents that repair work that is done as identified by preventive and predictive maintenance. It includes the inspection and servicing of jobs that are performed at specified recurring intervals to avoid more costly repairs and consequent downtime (preventive maintenance). Planned maintenance also includes that maintenance work that is initiated well in advance of equipment failure due to early detection of symptoms that imply equipment breakdown will take place in the near future unless servicing of worn or defective parts takes place (predictive maintenance). This planned maintenance will take full advantage of materials planning. Thus the materials, both stock and stores, required on maintenance or construction work is specified in advance. To perform this estimating accurately will usually require checking of the job site. When nonstock materials are needed alternative materials may be considered. Materials planning will permit more reliable scheduling and will increase the performance of maintenance crews by saving the time they spend by going to the job and then coming back to the storerooms and stockrooms to acquire the materials they need.

Job planning makes both daily and weekly scheduling possible. Regular feedback of information regarding the status of the repair job is an important integral of job planning that facilitates the scheduling procedure. Thus even emergency jobs can be planned in detail and be scheduled after the first day that work has begun.

Emergency Maintenance

Emergency maintenance, sometimes called critical maintenance, refers to any maintenance or construction job that should be started the same day that it is requested. Obviously, emergency maintenance usually cannot be introduced into a daily work schedule that promotes the most effective use of materials and manpower. The thorough planning that precedes scheduled maintenance is not practical in emergency maintenance because of the time factor. The amount of emergency maintenance that takes place should be very small when compared to the amount of scheduled maintenance (usually less than 10%).

There are two techniques for the handling of emergency maintenance.

1. Introducing the emergency maintenance into the regular maintenance schedule and then picking up the backlog with overtime, temporary workers, or outside contractors. (This is the preferred method.)
2. Assigning skilled craftsmen (one or more) whose sole responsibility is the handling of emergency work orders.

It is important that all work requests for emergency jobs be sent to a central point. These requests are usually telephoned in so that a work order can be written up and work begun immediately. A written work request should, of course, follow up the telephone communication.

As soon as work is begun and it is possible to estimate the extent of the repair of the emergency job, the remainder of the job should be planned and scheduled based upon the feedback information received.

Reliability Improvement

The reliability of a piece of production equipment is usually expressed as the mean time between failures or shutdowns (MTBS) in hours. The reliability, of course, has a significant impact on the equipment availability, which the maintenance function endeavors to improve. Availability may be expressed as

$$A = \frac{MTBS}{MTBS + MTD}$$

where MTD = mean downtime in hours.

The reader should recognize that the frequency of emergency maintenance is a function of the failure rate of those units which cause in-service failures of the system, and therefore emergency maintenance is a function of the reciprocal of the system's mean time between failures.

For every n operating hour of the system, there will be on the average n/MTBS emergency maintenance actions which will need to be performed. The time required to perform these repairs will vary both with the extent of the preventive maintenance performed and the components that cause the failure.

Modern engineering maintenance management includes an ongoing program of reliability improvement so as to continually improve the availability of all production equipment.

Cost Reduction

Another continuing program of a sound maintenance organization is one that is directed toward cost reduction by applying methods engineering. The methods engineer continually studies the way work has been and is being accomplished with the thought toward developing a better way to perform the maintenance, subjecting all the direct and indirect operations related to maintenance to close scrutiny in order to introduce improvements that will make work easier to perform and will allow work to be done in less time with less investment. Some considerations given in methods analysis include:

1. The maintenance materials used, the use of alternative materials, and standard replacement parts. Replace with more efficient equipment, for example, more efficient motors.
2. Inspection of the required equipment, on-site versus off-site inspection, inspection equipment available, and inspection procedure.
3. Maintenance repair and overhaul procedure, size of crew, crafts within crew, and procedure to be followed.
4. Materials handling on site and bringing materials to the site.

5. Tool equipment, use of power tools, and utilization of most efficient methods.

There are innumerable ways that maintenance can be performed; the methods engineer continually seeks a more cost-effective way.

Training and Skill Development

A sound maintenance program should include a training activity to help assure that the employees have the basic manual skills of the crafts involved, safety skills, craft judgment, skill in communicating technical information, and adequate flexibility so that diversified maintenance can be performed without complete dependence upon the specialist.

The training that takes place involves three classes: induction training, apprentice training, and modern techniques training.

Under induction training, the new maintenance worker is provided a short course to develop a feel for the job and the nature of the company's products. The employee becomes familiar with the reading of the company's prints so that he or she will be better able to communicate with sketches to others in the plant. These employees will also be given a brief exposure to the materials, tools, and equipment being used; the safety rules and regulations; and the nature of all production and inspection processes.

Apprentice training programs usually begin with ''helper'' training. New helpers should be carefully selected, giving prime consideration to potential employees who are interested in becoming craftsmen in time and who are capable of doing so.

Good judgment should be used in selecting the employees to whom the new helpers are assigned. Finally, helpers should be given a variety of experience. The actual training of the helper in learning all the skills of the basic craft along with the other qualities sought should be part of a formalized program.

A modern techniques program is one that periodically brings the latest maintenance techniques to all concerned. This program includes the utilization of equipment suppliers, staff engineers, and on occasion outside experts so that the latest thinking can be evaluated for possible application. In Chapter 15, training for maintenance work will be discussed in detail.

ISO 9000

ISO 9000 is a quality assurance management system that is rapidly becoming the world standard for quality. The ISO 9000 series standards are a set of four individual but related international standards on quality management and quality assurance with one set of application guidelines. These standards were developed in 1987 by the International Organization for Standardization, Geneva, Switzerland and have been adopted by the European Community. Today, more than one hundred countries, including the United States, have accepted ISO 9000 as the quality system of choice. The system incorporates a comprehensive review process, covering how companies design, produce, install, inspect, package, and market products. As a series of technical standards, ISO 9000 provides a three-way balance between internal audits, corrective actions, and corporate management participation leading to the successful implementation of sound quality procedures.

The series of technical standards include four divisions:

ISO 9001 is the broadest standard, covering procedures from purchasing to service of the sold product.
ISO 9002 is targeted towards standards related to processes and the assignment of subcontractors.
ISO 9003 sets technical standards that apply to final inspection and testing.
ISO 9004 sets standards that apply to quality management systems.

In order to become certified for one or more of the enumerated divisions, a company is obliged to produce documentation confirming that a positive situation exists in connection with those methods and procedures being screened. For example, Fig. 2.15 illustrates a preliminary profile of a fictious company seeking certification under ISO 9001. Note that considerable improvement is needed in process control, measuring equipment, design control, document control and purchasing.

In order to be considered a viable contender in the world market place, a manufacturer will need to be recognized as being certified under ISO 9000. Having a working engineering maintenance management program in place to assure the predictability of quality output by process equipment is important in connection with the certification of a manufacturing enterprise. Part of ongoing modern maintenance control systems include ISO 9000 certification.

By introducing an ISO 9000 program, leading to the building of quality excellence and eventual certification, an industry can anticipate

1. The lowering of manufacturing costs
2. A lower percentage of rejects
3. An overall improvement in product quality
4. Greater employee involvement

Estimates on representative manpower requirements for certification under ISO 9000 are listed in Table 2.4.

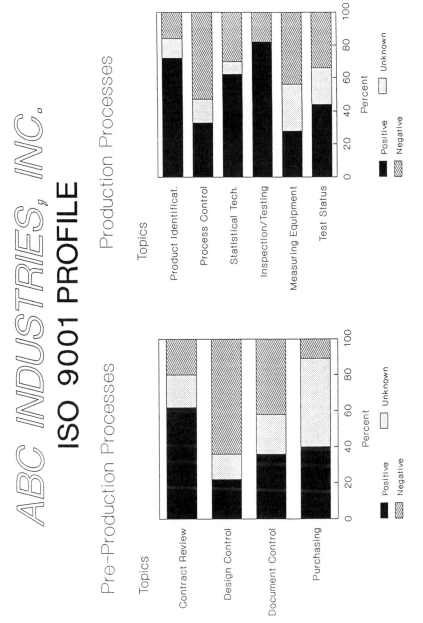

Figure 2.15 Preliminary profile of fictitious company seeking ISO 9001 certification.

Table 2.4 Estimates of Representative Manpower Requirements for Certification Under ISO 9000

	Man-days		
Items	ISO 9000	ISO 9002	ISO 9003
Contact and document			
Review	1	1	1
Preassessment visit	1	1	1
Quality audit	4–5	3–4	1–2
Report and certificate	1–2	1	1
Totals	7–9	6–7	4–5
Follow-up visit	1	1	1
Periodic audit	1–2	1	1
Triannual audit	3	2	1

The certification fee can vary considerably. It can be as low as $8000 or as high as $80,000. Considerable additional costs should be anticipated in connection with much introspection, internal analysis, and documentation resulting from implementations of procedures, controls, and standards.

U.S. firms that are heavily involved include Eastman Kodak, E. I. du Pont, AT&T, 3M, General Electric, Teledyne, and others. The accreditation body in the United States is the American Society for Quality Control (ASQC). The American Telephone and Telegraph Company serves as a registration agency.

SUMMARY

In order to maximize the impact of the engineering maintenance management program, there must be carefully planned control systems. Five maintenance management programs are emphasized. These are planned maintenance (predictive and preventive), emergency maintenance, reliability improvement, maintenance cost improvement, and training and skills development. For these five separate programs to function smoothly and be fully integrated with one another, a work order system should be established that provides the necessary communication to have the necessary work completed utilizing the best method so that the job can be completed on schedule. The work order system may be an entirely manual system or may be part of a data processing department. Following the issuance of informative work orders, a dispatching procedure needs to be followed. The maintenance department must maintain a scheduling and time reporting system so that work is performed as planned and management control information is gathered. This information should be reported in a manner that provides the basis for improvement. Improvement in performance can be estimated based upon the manufacturing progress function which includes productivity improvement realized by learning.

The growing need for quality improvement in order to be globally competitive involves increased performance from engineered maintenance. The international standard that includes an effective maintenance effort is ISO 9000. Certification under ISO 9000 is becoming increasingly prevalent throughout industry.

SELECTED BIBLIOGRAPHY

Mann, Lawrence, Jr. *Maintenance Management.* Lexington, MA: Heath, 1976.
Nakajima, Seiichi, Yamashina, Hatime, Kumagai, Chitoku, and Toyota, Toshio. "Maintenance Management and Control." In Gavriel Salvendy, Ed., *Handbook of Industrial Engineering*, 2nd ed. New York: Wiley, 1992.
Newbrough, E. T., and the staff of Albert Ramond and Associates, Inc. *Effective Maintenance Management.* New York: McGraw-Hill, 1967.
Niebel, B. W. *Motion and Time Study.* Homewood, IL: Richard D. Irwin, 1993.

3
Work Sampling in Maintenance

Work sampling is a technique used to investigate the proportions of total time devoted to the various activities that are comprised by a job or work situation. It is an especially powerful tool in studying maintenance work where there are a large number of interruptions and delays. It can identify areas of inefficiency and poor performance where methods analysis can be introduced and improvements made. In addition to locating problem areas, the results of work sampling are effective for determining allowances (personal and unavoidable) applicable to the job, for determining machine and man-power utilization, and for establishing standards of production.

In conducting a work sampling study, the analyst takes a comparatively large number of random observations at random intervals. The ratio of the number of observations of a given state of activity to the total number of observations taken will approximate the percentage of time that the process is in that given state of activity. For example, if 8000 observations at random intervals over a period of several weeks show that the first-shift crew of maintenance workers were working in 3600 instances and were not working in 4400 instances for a variety of reasons, then it would be reasonably certain that the expected productive working time of the maintenance crew was 45% and that 55% of the time was spent in such delays, interruptions, and related efforts, such as waiting for direction from supervisors, planning the job, looking for tools, repairing tools, waiting for other craftsmen, waiting for transportation, personal time, traveling to and from the job, traveling for tools, and so on.

The accuracy of the data determined by work sampling depends on the number of observations and the period over which the random observations are taken. Unless the sample size is of sufficient quantity, and the data are taken over a period of time that represents typical conditions, the results can not be reliable.

Much of the information acquired by work sampling can be obtained by conventional time study procedure. However, work sampling has some distinct advantages:

1. It does not require continuous observation by an analyst over a long period of time.
2. Clerical time is greatly diminished.
3. The total man-hours expended by the analyst are fewer.
4. The maintenance workers are not subjected to stopwatch observations for long periods.
5. Maintenance crews can be readily observed by a single analyst.

The theory of work sampling is based on fundamental laws of probability. If at a given instant an event can be either present or absent, statisticians have derived the following expression which shows the probability of x occurrences of an event in n observations.

$$(p + q)^n = 1$$

where

p = probability of a single occurrence

q = $(1 - p)$ the probability of an absence of occurrence

n = number of observations

If the preceding expression, $(p + q)^n = 1$, is expanded according to the binomial theorem. The first term of the expansion will give the probability that $x = 0$, the second term $x = 1$, and so on. The distribution of these probabilities is known as the binomial distribution. Statisticians have also shown that the mean of this distribution is equal to np and that the variance is equal to npq. The standard deviation is, of course, equal to the square root of the variance.

One may logically ask, ''Of what value is a distribution that allows only one event to either occur or not occur?'' For the answer to this, consider the possibility of taking one condition of the work sampling study at a time. All the other conditions can then be considered as nonoccurrences of this one event. Using this approach, we can now proceed with a discussion of binomial theory.

Elementary statistics tell us that as n becomes large, the binomial distribution approaches the normal distribution. Since work sampling studies involve quite large sample sizes, the normal distribution is the satisfactory approximation of the binomial distribution. Rather than use the binomial distribution, it is more convenient to use the distribution of a proportion with a mean of p, that is np/n, and a standard deviation of $\sqrt{pq/n}$, that is $\sqrt{npq}/n$, as the approximately normal distributed random variable.

In work sampling studies, we take a sample of size n in an attempt to estimate p. We know from elementary sampling theory that we cannot expect the $\hat{p}$ ($\hat{p}$ = the proportion based on a sample) of each sample to be the true value of p. We do, however, expect the $\hat{p}$ of any sample to fall within the range of $p \pm 2$ sigma approximately 95% of the time. In other words, if p is the true percentage of a given condition, we can expect the $\hat{p}$ of any sample to fall outside the limits $p \pm 2$ sigma only about 5 times in a hundred due to chance alone. This theory will be used to derive the total sample size required to give a certain degree of accuracy. It will also be used for the determination of subsample sizes.

ILLUSTRATIVE EXAMPLE

To clarify the fundamental theory of work sampling, we will interpret the results of an experiment. We have a production facility that has random downtimes. Let us take 8 random observations per day for a 100 day period. Then let

n = number of random observations per day

k = total number of days that observations are taken

N = total number of random observations

x_i = number of down observations in n random observations on day i ($i = 1, 2, \ldots, k$)

N_x = number of days that the study showed x downs in n random observations ($x = 0, 1, 2, \ldots, n$)

$P(x)$ = probability of x observations down in n observations based on the binomial distribution

$$P(x) = \frac{n!}{x!(n - x)!} p^x q^{n - x}$$

p = the probability of the production facility being down

q = the probability of the production facility running

$p + q = 1$

$\bar{P}_i = x_i/n$, observed proportion of downtime on day i ($i = 1, 2, 3, \ldots, k$)

$\hat{P} = \sum_{i=1}^{k} x_i /N$, expected proportion of downtime for entire study

Let us assume that an all-day time study for several days revealed that $p = 0.5$. Table 3.1 shows the number of days in which x breakdowns were observed ($x = 0, 1, 2, 3, \ldots, n$) and the expected number of breakdowns given by our binomial model.

The reader can observe the close agreement between the observed number of days that a specified number of breakdowns were encountered (N_x) and the expected number computed theoretically $100p(x)$ when $p = 0.5$.

In the typical industrial situation, p (which was observed in our example to have a value of 0.5) is unknown to the analyst. The best estimate of p is $\hat{p}$, which may be computed as x/n. As the number of observations taken at random per day (n) increases, $\hat{p}$ will approach p. However, with a limited number of random observations, the analyst is concerned with the accuracy of $\hat{p}$.

Table 3.1 Actual and Expected Number of Days with 0 to 8 Breakdowns

x	N_x	$P(x)$	$100P(x)$
0	0	0.0039	0.39
1	4	0.0312	3.12
2	11	0.1050	10.5
3	23	0.2190	21.9
4	27	0.2730	27.3
5	22	0.2190	21.9
6	10	0.1050	10.5
7	3	0.0312	3.12
8	0	0.0039	0.39
	100	1.001[a]	100[a]

[a]Approximately.

If a plot of $p(x)$ versus x were made from the preceding example, it would appear as illustrated in Fig. 3.1.

When n is sufficiently large, regardless of the actual value of p, the binomial distribution will very closely approximate the normal distribution. This tendency can be seen in the above example when p is approximately 0.5. When p is near 0.5, n may be small and the normal can be a good approximation to the binomial. When using the normal approximation, we set the mean u equal to p and the standard deviation $\sigma_p = \sqrt{pq/n}$.

To approximate the binomial distribution the variable z used for entry in the normal distribution may take the following form:

$$z = \frac{(\hat{p} - p)}{\sqrt{pq/n}}$$

Although p is unknown, in the practical case we can estimate p from $\hat{p}$ and can determine the interval within which p lies using confidence limits. For example, we can imagine that the interval defined: $\hat{p} - 2\sqrt{\hat{p}\hat{q}/n}$ and $\hat{p} + 2\sqrt{\hat{p}\hat{q}/n}$ contains p 95% of the time. Graphically, this may be expressed as

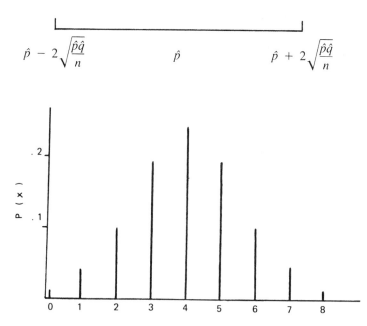

Figure 3.1 Plot of $P(x)$ versus x.

We can derive the expression for finding a confidence interval for p as follows. Let us suppose that we want an interval which will contain p 95% of the time, that is, a 95% confidence interval. For n sufficiently large, the expression

$$z = \frac{\hat{p} - p}{\sqrt{\hat{p}\hat{q}/n}}$$

is approximately a standard normal variable. Therefore, we can set the probability

$$p\left(z_{0.025} < \frac{\hat{p} - p}{\sqrt{\hat{p}\hat{q}/n}} < z_{0.975}\right) = 0.95$$

Rearranging the inequalities then gives

$$p\left(\hat{p} - z_{0.975}\sqrt{\frac{\hat{p}\hat{q}}{n}} < p < \hat{p} + z_{0.975}\sqrt{\frac{\hat{p}\hat{q}}{n}}\right) = 0.95$$

Remember that $-z_{0.025} = z_{0.975} = 1.96$ or approximately 2. The interval with approximately a 95% chance of containing p is then

$$\hat{p} - 2\sqrt{\frac{\hat{p}\hat{q}}{n}} < p < \hat{p} + 2\sqrt{\frac{\hat{p}\hat{q}}{n}}$$

These limits imply that the interval defined contains p with 95% confidence since z has been selected as having a value of 2.

The underlying assumptions of the binomial are that p, the probability of a success (the occurrence of downtime), is a constant each random instant that we observe the process. Therefore, it is always necessary to take random observations when taking a work sampling study.

PLANNING THE WORK SAMPLING STUDY

Before beginning work sampling observations, the study should be carefully planned and explained to those foremen in the maintenance department where the work sampling study or studies will be conducted.

The first step is to make a preliminary estimate of the activities on which information is being sought. Often this estimate may be acquired from historical data. If such data does not exist, the analyst can work sample the area or areas involved for a short period (two or three days) and use the information obtained as the basis of estimates. Once the preliminary estimates have been made, the analyst should determine the accuracy that is desired of the results. This can best be expressed as a tolerance within a stated confidence level.

Having made a preliminary estimate of the percentage occurrences of the elements being studied, and having determined an accuracy desired at a given confidence level, the analyst will now make an estimate of the number of observations to be made. Knowing how many observations need to be taken and the time that is available to conduct the study, the frequency of the observations to be taken can be determined.

The next step is to design the work sampling form or card on which the data will be tabulated and the control charts that will be used in conjunction with the study.

DETERMINING THE OBSERVATIONS NEEDED

In order to determine the number of observations needed, the analyst must first estimate how accurate his results should be. The larger the number of the observations, the more valid the final answer will be. Four thousand observations will give considerably more accurate results than will 400. However, if the accuracy of the result is not the prime consideration, 400 observations may be ample.

In random sampling procedures, there is always the chance that the final result of the observations will be beyond the acceptable tolerance. However, sampling areas will diminish as the size of the sample increases. The standard error σ_p of a sample proportion or percentage as shown in most textbooks on statistics may be expressed by the equation

$$\sigma_p = \sqrt{\frac{pq}{n}} = \sqrt{\frac{p(1-p)}{n}}$$

where

σ_p = standard deviation of a percentage

p = true percentage of the element being sought expressed as a decimal

n = total number of random observations upon which p is based

By estimating the true percentage occurrence of the element being sought (which we can designate as $\hat{p}$) and by knowing the standard error, it is possible to substitute in the above expression and compute n.

$$n = \frac{\hat{p}(1-\hat{p})}{\sigma_p^2}$$

For example, it is desired to determine the number of observations required with 95% confidence so that the true proportion of unavoidable delay time for our electricians is within the interval 10–18%. It is anticipated that the unavoidable delay time encountered by electricians is about 14%. These assumptions are illustrated graphically in Fig. 3.2.

In this case, $\hat{p}$ would equal 0.14 and $\sigma_{\hat{p}}$ would equal 0.02. Solving for n,

$$n = \frac{0.14\ (1 - 0.14)}{(0.02)^2} = 301 \text{ observations}$$

After the initial estimate of the number of observations has been obtained, a more accurate estimation of p may be computed by recalculating n based on the values of $\hat{p}$ and σ_p calculated from the first few days of observations. If the desired accuracy has not been obtained more observations are taken and the above process is repeated.

DETERMINING THE FREQUENCY OF THE OBSERVATIONS

The frequency of the observations depends primarily on the number of observations required and the time limit placed on the development of the data. For example, if 1600 observations were needed and the study needed to be completed in the next 40 working days, it would be necessary to obtain approximately

$$\frac{1600 \text{ observations}}{40 \text{ working days}} = 40 \text{ observations per working day.}$$

Let us assume 5 electricians are involved in the study. The analyst would then need to make 8 random trips per day to determine the exact activity of each electrician at the precise random time he is observed. The time of day selected for these 8 observations should be chosen at random daily. Thus, no set pattern should be established from day to day for the time when the analyst appears on the production floor.

A simple technique that may be used is to select 8 numbers daily from a statistical table of random numbers. Assuming the table of random numbers

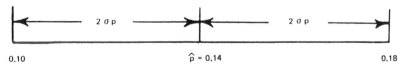

0.10 $\hat{p}$ = 0.14   0.18

Figure 3.2 The tolerance range of the percentage of unavoidable delay allowance for electricians.

is 1 through 100 and that the workday is 480 minutes (8 hours), we can let a unit of each random number equal 4.8 minutes. Thus the number 10 would be equivalent to 10 × 4.8 or 48 minutes, and the analyst would make a trip to observe the five electricians at 48 minutes after the beginning of the work shift. Each day of the study, the analyst in this way can determine the exact time of day a trip to the work stations will be made to observe the activity of each electrician.

The microcomputer can also be used advantageously to determine the schedule of daily observations. Computer programs are available (or may easily be developed) to provide the time of day to take each of a series of random observations of the work station. For example, one program that has been developed is based upon a 24-hour clock so that random observations may be taken at any time over three 8-hour shifts. The input data only require the time that the shift under study begins and ends and the number of random observations to be taken during the shift.

DESIGNING THE WORK SAMPLING FORM

It is a good idea to design an observation form to best record the data that will be gathered during the course of the work sampling study. Standard forms usually are not completely appropriate since each work sampling study is unique from the standpoints of the information being sought, the random times that observations will be made, and the total number of observations needed. The best form is one that is tailored to the objectives of the study.

A work sampling form is illustrated in Fig. 3.3. This form was designed to determine the time being utilized for various productive and nonproductive states in a maintenance repair shop. This form was designed for up to twenty random trips to the main repair shop per day.

USE OF CONTROL CHARTS

The control chart techniques used so extensively in statistical quality control work can be applied readily to work sampling studies. Since work sampling studies deal exclusively with percentages or proportions, the "p" chart is used most frequently.

The first problem encountered in setting up a control chart is the choice of limits. In general, a balance is sought between the cost of looking for an assignable cause when none is present and not looking for an assignable cause when one is present. The three-sigma limit is generally used for establishing control limits on the "p" chart.

WORK SAMPLING STUDY

Main Repair Shop _____

Number Working This Study _____ Date _____   By _____

Remarks _____

Obs. No.	Random Time	Productive Occurrences							Nonproductive Occurrences							Total Observations	Percentage Productive	Percentage Nonproductive
		Mch	Weld	Pipe Fit	Gen. Labor	Elect.	Carpen.	Janitor	Get Tools	Grind Tools	Wait Job	Wait Crane	Confer Foreman	Personal	Idle			
1																		
2																		
3																		
4																		
5																		
6																		
7																		
8																		
9																		
10																		
11																		
12																		
13																		
14																		
15																		
16																		
17																		
18																		
19																		
20																		
Totals																		

Figure 3.3 Work sampling study.

Suppose that "p" for a given condition is 0.10 and 180 samples are taken each day. By substituting in the standard error equation for p and n, control limits of ± 0.07 are obtained.

$$
\begin{aligned}
\sigma_p &= \sqrt{\frac{\hat{p}(1 - \hat{p})}{n}} \\
&= \sqrt{\frac{(0.10)(1 - 0.10)}{180}} \\
&= 0.0223607 \\
\pm 3\sigma_p &= (3)(0.0223607) \\
&= 0.07
\end{aligned}
$$

A control chart similar to Figure 3.4 can then be constructed. The p' values for each day would be plotted on the chart. (Note that $\Sigma p'/N = \hat{p}$, where N equals the number of days over which the work sampling study was made, with approximately the same number of observations made each day of the study.)

The analyst in work sampling considers points beyond the three-sigma limits of p as out of control. Thus a certain sample that yields a value of p' is assumed to have been drawn from a population with an expected value of p if p' falls within the plus-or-minus three-sigma limits of p. Expressed another way, if a sample has a value of p' that falls outside the three-sigma limits, it is assumed that the sample is from some different population or that the original population has been changed.

Points other than those out of control may be of some statistical significance. For example, it is more likely that a point will fall outside the three-sigma limits than that two successive points will fall between the two- and three-sigma limits. Hence, two successive points between the two- and three-sigma limits would indicate that the population has changed. Series of significant sets of points have been derived. This idea is discussed in most statistical quality control texts under the heading "Theory of Runs."

A hypothetical example will show how control charts can facilitate a work sampling study. The reliability engineer wishes to measure the percentage of production machine downtime that is occurring in the turret lathe department. An estimate based on two days of sampling indicates this downtime to be approximately 0.20. The desired results are to be within plus or minus 5% of p with a level of significance of 0.95. The sample size is computed to be 6400. It was decided to take the 6400 readings over a period of 16 days at the rate of 400 readings per day. A p' chart was computed for each daily sample of 400. A p chart was set up for $p = 0.20$ and subsample size $n' =$

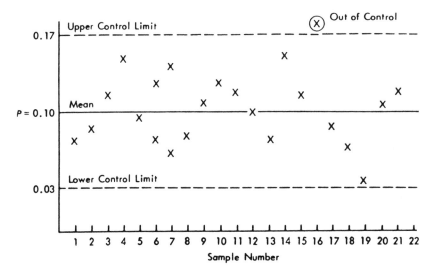

Figure 3.4 Control chart.

400. Each day readings were taken and p' was plotted. On the third day, the point for p' went above the upper control limit. An investigation revealed that there had been an accident in the lathe section of the plant and that several of the men had left their machines to assist the injured employee to the plant hospital. Since an assignable cause of error was discovered, this point (p' for the third day) was discarded from the study. If a control chart had not been used, these observations would have been included in the final estimate of p.

On the fourth day, the point for p' fell below the lower control limit. No assignable cause could be found for this occurrence. The reliability engineer in charge of the project also noted that the p' values for the first two days were below the mean p. Consequently, he decided to compute a new value for p using the values from days one, two, and four. The new estimate of p turned out to be 0.15. To obtain the desired accuracy, n is now 8830 observations. The control limits also change as shown in Fig. 3.5. Observations were taken for 12 more days and the individual p' values were plotted on the new chart. As can be seen, all the points fell within the control limits. A more accurate value of p was then calculated using all 6000 observations. The new estimate of p was determined to be 0.14. A recalculation of achieved accuracy showed it to be slightly better than the desired accuracy. As a final check, new control limits were computed using p equal to 0.14. The dashed lines shown in Fig. 3.5 indicate that all points were still in control using the new

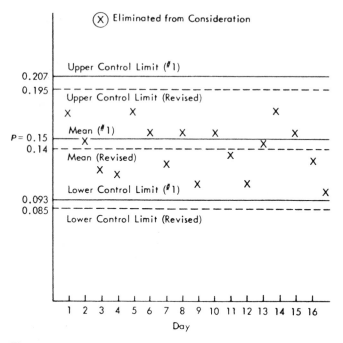

Figure 3.5   Control limits can change.

limits. If a point had fallen out of control, it would have been eliminated and a new value of p would have been computed. This process would then have been repeated until the desired accuracy was achieved and all p' values were in control.

The reader should understand that although the percentage downtime in this turret lathe department is 14% today, it should not necessarily be 14% in the future. Improvement should be a continuing process and consequently downtime should diminish. Reduction of downtime is a prime objective of a sound maintenance program. The principal purpose of work sampling is to determine areas of direct and indirect work that might be improved.

OBSERVING AND RECORDING THE DATA

A representative sample form for a shift study in a maintenance repair shop is shown in Fig. 3.6. Here 6 random observations of each facility were made per shift. A digit designating the particular observation was indicated in the

WORK SAMPLING OBSERVATION RECORD

STUDY NO. 38-84
PLANT Reading
DATE 8/14
OBSERVER G. Thuering

STUDY AREA Main Repair Shop

REMARKS

MACHINE	DRAWING	CUTTING	SET-UP	MACHINE IDLE	CRANE WAIT	WAIT INSPECTION	AID INSPECTION	WAIT-TOOLS NOT AVAILABLE	WAIT-TOOL TROUBLE	CONFER WITH ENGINEER	CONFER WITH FOREMAN, INSP.	TOOL HANDLING	GET OR GRIND TOOLS	REMOVE CHIPS, CLEAN WORK STATION	WAIT FOR JOB
Hardinge 11"	98 L 16701	3	1	2,4,6											
Leblond Making 19"	75 K 22403	1,2,4,5								5	3		6		
DOAll 35"	78 K 34161	6	4	1,2,3					5				6		
Bridgeport CNC	17 J 14946	2,5	6	3								1			
Cincinnati Milacron	45 L 32614	1,4				5	2								
Ex-Cell-O 3-axis milling	98 L 16744	3,5		1,2				4		6	5	6			
Ex-Cell-O Honing	35 M 38414	3,4		1,2					3					5	
Webster Bennett Vert.	81 K 42112	4,6	3		1		2							5	
Jones & Shipman Surface Grinder	14 J 97211	1,6		2,3	4										
DoAll Contour	73 K 86419	2,5,6	1	3,4									5		
Kysor Horizontal Band Saw	16 K 77312	1,2,4,6 3	3									5			
Apex Horizontal Broach	84 L 16912	5		1,2,3,4,6											5
Cincinnati Radical 2' Drill	76 J 84410	1,4,6	5	3,4											
L.G. Single Spindle Drill	65 K 13888	2,3,4		1,6										5	
		84 = 32	+ 7	+ 24	+ 1	+ 2	+ 1	+ 2	+ 1	+ 2	+ 3	+ 2	+ 3	+ 3	+ 2 + 2

Figure 3.6 Typical work sampling form.

61

space provided for the state of each facility under study. Since 14 facilities were being studied, a total of 84 observations were made per shift.

The analyst should not anticipate an expected recording while approaching the work area. Instead, a point at a given distance from the facility should be used for making the observation and recording the facts. It might be helpful to make an actual mark on the floor to show where the analyst should stand for making observations. If the operator of the machine being studied is idle, the analyst should determine the reason for idleness, confirming the reason with the line foreman, before making the proper entry. The analyst should learn to take visual observations, making written entries after leaving the scene of the work area. This will minimize the feeling of being watched on the part of the shopworkers and they will perform in their accustomed manner.

To help assure that the workers perform in their usual fashion, it is advisable to inform them of the purpose of the study. The fact that no watch is used tends to relieve the operators of a certain mental tension; little difficulty is experienced in getting their full cooperation.

SUMMARY

Work sampling allows the analyst to get important facts easily and reasonably soon. For example, Table 3.2 illustrates the information gathered by the work sampling technique in the maintenance department of a medium-sized electrical appliance firm. Note this detailed tabulation shows the difference between one labor group and another with a range in working time from 45.1% to 83.5%. This study was able to point out several areas that warranted investigation. It provided a basis of comparison for the maintenance operation in one plant with that of its sister plants. And, most important, these results provided the data for better planning, scheduling, and supervision within the maintenance department.

Performance-rated work sampling is especially useful in determining the amount of time that should be allocated for unavoidable delays, work stoppages, etc. The extent of these interruptions is a suitable area for study to improve productivity.

Today many companies are using computers to make the repetitive calculations characteristic of work sampling. The computer can be used to calculate not only daily results, but also cumulative results, such as the maintaining of control charts. Thus automation of the clerical work—including the recording of observations; the computation of element percentages, performance ratings, and statistical accuracies; the preparation and maintenance of control charts;

Table 3.2   Typical Information Gathered by Work Sampling

Labor Group	Working	Getting materials	Clean up/get tools	Timekeeping	Instruction	Traveling	Waiting	Not working	Personal	No contact
Electricians	55.5	3.2	2.8	1.0	4.0	18.0	0.5	5.3	0.8	8.8
Carpenters	71.6	1.3	3.0	0.8	2.2	10.3	0.4	7.6	0.5	2.4
Millwrights	66.0	1.4	2.2	0.7	5.0	12.0	1.3	5.5	0.9	5.0
Pipe fitters	52.9	1.5	5.6	1.5	3.2	17.2	1.9	4.7	0.3	11.3
Foundry	45.1	3.4	5.9	2.2	4.0	20.8	1.2	6.8	1.9	8.4
Laborers	83.5	—	0.8	0.3	0.3	7.5	1.4	5.3	0.3	1.9
Storekeepers	81.1	4.5	0.6	0	0.6	6.4	1.3	5.1	—	—
Average	61.7	2.0	3.1	1.2	3.5	14.4	1.2	5.4	0.7	6.8

and the extrapolation of the data into equivalent manpower or machines and annual costs—can take place with the use of the computer.

SELECTED BIBLIOGRAPHY

Barnes, R. M. *Work Sampling*, 2nd ed. New York: Wiley, 1957.

Heiland, R. E., and Richardson, W. J. *Work Sampling*. New York: McGraw-Hill, 1957.

Niebel, B. W. *Motion and Time Study*, 9th ed. Homewood, IL: Richard D. Irwin, 1993.

Pape, E. S. Work Activity Sampling—Contemporary Design Analysis Methodology and Applications, Part II: Work Sampling Calculations Revisited. American Institute of Industrial Engineers, 1979 Fall Industrial Engineering Conference Proceedings, Norcross, GA, 1979.

Richardson, W. J. *Cost Improvement, Work Sampling and Short Interval Scheduling*. Reston, VA: Reston Publishing, 1976.

4
Estimating and Measuring Maintenance Work

The desirability of establishing accurate standards on maintenance work economically should be apparent. Good standards are needed for planning, scheduling, and cost control. And since maintenance continually requires a greater proportion of the total cost of operation of an enterprise, it is important that the maintenance function become as efficient as possible.

In most enterprises, the development of sound standards in maintenance and other indirect work has not been as effective as in the establishment of standards on direct work. This is primarily due to three reasons.

1. Maintenance work, on the average, has a much longer cycle than direct labor operations.
2. A large proportion of maintenance is nonrepetitive. Although there is a great deal of similarity between one job and the next, there is a variation in content from job to job.
3. Working conditions vary considerably more in maintenance work than in direct labor operations.

Reliable standards can be established in advance (during the planning stage) of maintenance work using techniques based on measurement. These techniques include standard data, formulas, and slotting procedures. Too often, time shown on work orders is based solely upon approximations pro-vided by the maintenance planner. Although rough estimates are better than providing no information on the time anticipated to be taken by each operation of the maintenance work order, experience has proven that estimates based upon judgment alone are seldom accurate. By providing reliable standards based upon sound work methods, the maintenance analysts will find dramatic cost improvements while ensuring quality work.

Rough estimates are seldom accurate for several reasons. First, the individ-ual doing the estimating seldom has significant work experience in all facets of the work being estimated. Second, there is a natural tendency for the planner to estimate high in order to assure that both line supervision and workers will have adequate time to perform the work and, consequently, will not be critical of the estimator. Third, the estimator usually will not take the time to utilize the best available estimating methods, which necessitate break-ing the work down in elements and then estimating the time for each individual element. Fourth, most estimates are based upon historical records which seldom are valid for all work that is to be performed in the future.

The reader is urged to insist that a standards program for all maintenance work be based upon one or a combination of these three methods: standard data, formulas, or slotting techniques. Standards should not be based upon the planner's approximations. Companies that fully utilize reliable standards based on measurement have proven conclusively the cost-effectiveness of the procedure. It is very easy for the inexperienced analyst introducing a maintenance management system to forego the slow, tedious experience of building a statistical standard data bank from which maintenance standards data and formulas can be developed. It may seem more prudent to utilize the experience of an intelligent maintenance employee to estimate the time of each job and record this standard on the work order. But this author wants to assure the reader that although this approximation procedure can be introduced much more rapidly, in a short time it will lead to the failure of the entire installation.

A simple test will confirm the unreliability of gross estimates. Take a very simple assembly such as a U-bolt, clamp, and two nuts; and have several workers (all with similar education level and training) independently estimate the standard time to assemble these four parts. There will likely be a variation in the estimates amounting to several hundred percent and probably no one person will estimate a time within ± 10% of the correct standard based upon measurement.

Since we want an accurate estimate of a fair standard for each operation shown on the work order in advance of the work being done, time does not permit using stopwatch techniques. However, both stopwatch studies and fundamental motion data will be needed in the development of standard data, formulas, and benchmark standards for use in slotting techniques. Thus, it is essential that both methods continually take place in conjunction with the engineering maintenance management effort.

STOPWATCH TIME STUDY

Time Study Equipment

The minimum equipment required to carry on a stopwatch time study program includes a stopwatch, a time study board, and an electronic calculator.

Stopwatches are available that are either mechanically operated or battery operated. They can be procured where time is measured in either 0.01 minute or 0.0001 hour. The mechanically operated decimal minute watch tends to be a favorite with work measurement analysts who develop standard data for maintenance work.

The time study form will carry all details of the study. It should provide space to record the pertinent information related to the method being employed. This includes details as to tools employed and their location; materials involved and their condition; information on feeds, speeds, and depths of cut; inspection requirements; and distances traveled. The operation being performed should be completely identified as to what is being done on what facility. It is always better to provide too much rather than too little information concerning the job being studied.

Of course the form should be designed so that the analyst can conveniently record his watch readings and performance rate, the elemental times. The form should also allow space for the working up of the study. Figure 4.1 illustrates a time study form that has been designed for maintenance work. In this form, the various elements of the operation are recorded horizontally across the top of the sheet, and the cycles studied are recorded vertically, row by row. The one-page form will accommodate twenty elements and up to four cycles. In the vast majority of maintenance and repair studies only one cycle will occur on the majority of the elements.

The reading column has been divided into two sections. The large area headed "R" is for recording the watch reading, and in the small section marked "F" is shown the element performance factor. The "T" column is provided for elemental elapsed time values.

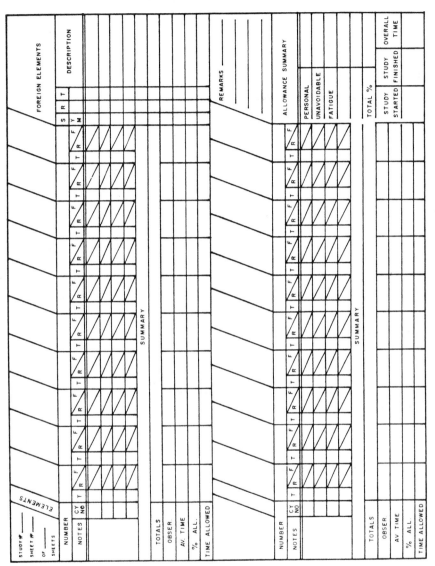

Figure 4.1a Time study form designed for maintenance work (front).

STUDY # _____ DATE _____ DEPT. WHERE PERFORMED _____

COMPLETE DESCRIPTION OF WORK PERFORMED

_____
_____
_____
_____

EQUIPMENT REPAIRED _____

FACILITIES USED _____
_____

MATERIALS USED _____
_____

CONDITIONS _____
_____

DWG. NOS. _____

DISTANCE TRAVELLED TO ACQUIRE TOOLS _____ MATERIALS _____ WORK _____

ELEMENT #	ELEMENT DETAILS — SMALL TOOL NOS., FEEDS, SPEEDS, ETC.	ELEMENTAL TIME	OCC. PER CYCLE	TOTAL TIME ALLOWED

TOTAL TIME _____ HRS _____

FOREMAN _____ INSPECTOR _____

OBSERVER _____ APPROVED BY _____

Figure 4.1b Time study form designed for maintenance work (back).

68

Approach to Operator's and Observer's Position

Before the start of the time study, the analyst should meet with the supervisor of the operation to be studied. When the supervisor understands the purpose and method of the study, the analyst should be introduced to the maintenance worker(s) involved. It is not advisable to study any operation without the prior knowledge of the responsible supervisor.

The maintenance worker should be approached in a friendly manner, be informed that the operation is to be studied, and should then be given the opportunity to ask any questions relative to such matters as the timing technique, method of rating, and application of allowances. In some instances the maintenance worker may have never been studied before and the time study analyst will find it worthwhile to answer the worker's questions frankly and patiently.

The time study analyst should show interest in the maintenance worker's job and at all times be fair and straightforward in his or her behavior to the worker.

As the maintenance worker begins the work under study, the time study observer should take a position a few steps to the rear of the worker to avoid distraction or interference. It is important that the time study observer stand while taking the study. During the course of the study, the observer should avoid any conversation with the worker as this would tend to upset the routine of both the analyst and the maintenance worker.

Dividing the Job into Elements

For future use as standard data and for ease of measurement, the maintenance job is divided into segments of work known as elements. Since the operation, in most cases, will not be repeated the observer will need to identify the elements as they are taking place, measure their duration, and record the facts during the course of the study.

Each element should be recorded in its proper sequence and should include a logical termination. Thus the first element might be "walk to tool crib (125 feet) and check out electrical repair tool kit"; the second element might be "walk from tool crib to #2 centerless grinder in main machine shop"; the third element might be "remove housing from #2 centerless grinder power switch." Some general considerations in breaking the job down into elements in maintenance work include the following:

1. Keep manual and machine time separate.
2. Select elements so that they can be readily and accurately timed.

3. Keep elements associated with planning, material acquisition, material transportation, and the actual repair or maintenance work separate.
4. Be sure that all elements being performed are necessary.

Taking the Study

There are two techniques of recording the elemental time while taking the study. The continuous method, as the name implies, allows the stopwatch to run for the entire duration of the study. In this method, the watch is read at the breaking point of each element while the hands of the watch are moving. A stopped hand is read under the continuous method when using a double-action watch. In the snapback technique, the watch is read at the termination point of each element and then the hands are snapped back to zero. As the next element takes place, the hands move from zero. The elapsed time is read directly from the watch at the end of this element, and the hands are again returned to zero. This procedure is followed throughout the entire study.

When recording the watch readings, the analyst notes only the necessary digits and omits the decimal point, thus giving as much time as possible to observing the performance of the workman. It is necessary to evaluate the performance of each element as it occurs during the study.

Since the actual time that was required to perform each manual element of study was dependent to a high degree upon the skill and effort of the worker, it is necessary to adjust the time of the good maintenance worker up to normal and the time of the poor worker down to normal.

In the performance rating or leveling system, the observer evaluates the effectiveness of the operator in terms of his conception of a "normal" operator performing the same element. This effectiveness is given a value expressed as a decimal or percentage and assigned to the element observed. A normal maintenance workman is defined as a qualified, thoroughly experienced worker who performs at a pace neither too fast nor too slow but representative of average. The reader should recognize the normal workperson exists only in the mind of the time study analyst, and the concept has been developed as a result of thorough and exacting training and experience in the technique of measuring wide varieties of work.

Applying Allowances

There are three classes of interruptions that take place occasionally for which extra time must be provided. These are

1. Personal interruptions for an employee's well-being, such as trips to the drinking fountain and rest room.

2. Unavoidable delays such as interruptions by the foreman or other management personnel, tool breakage, etc.
3. Fatigue. This is a natural slowing down of the operator as the workday progresses. The amount of fatigue that takes place is dependent upon the nature of the work as well as the working environment.

Since the time study is taken over a relatively short period and since foreign elements have been removed in determining the normal time, an allowance must be added to the leveled or base time to arrive at a fair standard that can be achieved by the normal maintenance worker when exerting average effort. Typical allowances for maintenance work range between 15% and 20%. However, at times, allowances of more than 20% need to be added to the normal time in order to compute a fair standard. On occasion maintenance work is performed under less than ideal working conditions that cause not only high fatigue but necessitate many unavoidable delays. For example, maintenance work may be performed under high or low temperatures or in wet or high wind conditions. Work can also be performed at heights or underground, which requires additional allowances to accommodate unusual fatigue and unavoidable delays. Table 4.1 provides the effect of working conditions in order to arrive at an allowance factor for personal delays and fatigue for maintenance work.

Calculating the Standard

Once the analyst has properly recorded all the necessary information on the time study form, recorded the elemental times, and properly performance-rated each element of the study, he should thank the operator and proceed to the next step which is computation of the study. The initial step in computing the study is to verify the final stopwatch reading with the overall elapsed clock reading. This check can only be made if the continuous method of watch reading was used. These two time values should check within a half-minute.

When using the continuous method, a subtraction is made between the watch reading and the preceding reading, giving the elapsed time. This procedure is followed throughout the study, each reading being subtracted from the succeeding one. The elapsed elemental times, after being multiplied by the leveling factor, are recorded in ink or red pencil in the T column of the time study form.

After the normal elemental times have been calculated, the percentage allowance is added to each element to determine the allowed elemental time. It is these normal and allowed elemental times that form the basis of standard data.

Table 4.1 Maintenance Allowances for Personal Delays and Fatigue

A. Constant Allowances	
1. Personal allowances	5%
2. Basic fatigue allowance	4%
B. Variable Allowances	
1. Must work while standing	2%
2. Must work in awkward position	
a. Considerable bending	2%
b. Requires stretching, lying	6%
3. Requires use of force or muscular energy	
a. Weight handled is 10 to 15 lb	1%
b. Weight handled is 15 to 20 lb	2%
c. Weight handled is 20 to 25 lb	3%
d. Weight handled is 25 to 30 lb	4%
e. Weight handled is 30 to 35 lb	5%
4. Work performed under poor visibility conditions	
a. Well below recommended light conditions	2%
b. Light is definitely inadequate	5%
5. Work performed under unfavorable temperature and/or humidity	
a. Work performed under hot or cold conditions	5%
b. Work performed under excessive heat, humidity, cold conditions, or high wind or wet conditions	8%
6. Work requires very close attention	
a. Fine or exacting requirements	2%
7. Noise level	
a. Work performed under intermittent and loud conditions	2%
b. Work performed at intermittent and very loud or high-pitched and loud conditions	5%

This standard data represents the information that is placed in the data bank of the computer that will be used for establishing reliable estimates for future work. Thus, the primary purpose of time study in connection with maintenance work is not to develop a standard for a given job, but to develop meaningful data that will be used to accurately estimate maintenance work performed in the future.

To determine a standard for the maintenance job being time studied, the analyst summarizes the allowed elemental values on the reverse side of the time study form. This standard can then serve as an audit of the standard established when the work order was initiated.

STANDARD DATA

Standard time data are elemental standard times taken from time studies and standard elemental times developed from fundamental motion data which will be discussed later. Standard data can exist in tabular form, curves, nomograms, and formulae.

Maintenance work standards calculated from standard data will be relatively consistant since the tabulated elements are the result of precise measurement of accepted methods used in the development of maintenance work standards. It is necessary to summarize only the appropriate tabulated elemental values to arrive at a fair standard. It should be apparent that standards on maintenance work can be computed relatively rapidly using the standard data technique. Table 4.2 illustrates typical standard data elements used in the development of a standard for the servicing of a 10-ton-capacity jack.

In addition to elemental standard data, operational standard data can be tabulated for those maintenance jobs that occur regularly such as planned maintenance work characteristic of predictive and preventive maintenance. For example, the following nine operations may be considered standard maintenance procedure every three months for a line of Brown and Sharpe standard grinders located in the main machine shop.

1. Grinding spindle. Check for wear on standard grinders. On universal grinders, adjust for end play after spindle is up to temperature.
2. Wheel and main slide. Check for wear.
3. Headstock and tailstock spindle. Check for wear.
4. Transmission. Check oscillating mechanism for wear.
5. Feed wheel. Check end play.
6. Infeed slide. Adjust backlash.
7. Belts. Check for condition. Replace if necessary
8. Coolant pump. Check operation and for wear.
9. Lubrication pump. Check condition.
   Allowed standard hours = 2.25

This 2.25 standard hours was developed by summarizing the standard times for the nine service operations listed where the standard times were developed either by stopwatch time study or by adding up the appropriate elemental standard data.

Maintenance standard data should be indexed and filed by facility. Typical maintenance standard data would be tabulated as follows:

1. Tool acquisition
2. Materials acquisition

Table 4.2 Elemental Standard Data Required to Disassemble and Reassemble
Hydraulic Plunger Cups in Portable 10-Ton-Capacity Jack (Includes Cleaning
and Oiling)

No.	Operation	Unit time (hr)	No. units	Total time (hr)
1.	Remove and reinstall two drive screws	0.011	4	0.044
2.	Remove and reinstall name plate	0.003	2	0.006
3.	Remove and reinstall spring retaining screw	0.011	2	0.022
4.	Remove and reinstall thread protector	0.016	2	0.032
5.	Remove and reinstall pin	0.023	2	0.046
6.	Remove and reinstall saddle	0.016	2	0.032
7.	Remove and reinstall plunger stop ring	0.016	2	0.032
8.	Remove and reinstall plunger ring	0.011	2	0.022
9.	Remove and reinstall spring adjusting screw assembly	0.011	2	0.022
10.	Remove and reinstall pull spring	0.023	2	0.046
11.	Remove and reinstall plunger cup retaining nut	0.016	2	0.032
12.	Remove and reinstall spring washer, spreader, plunger cups (3), cup separators (2), and disc	0.012	16	0.192
13.	Remove and reinstall plunger	0.012	2	0.024
14.	Clean 18 small parts	0.016	18	0.288
15.	Clean 2 medium parts	0.041	2	0.084
16.	Lubricate 12 small parts	0.007	12	0.084
Total				1.01

Source: Data adapted from Engineered Performance Standards, Department of the Navy,
Bureau of Yards and Docks.

  3. Machine or facility
      a. Constants
      b. Variables

Under tool acquisition, data is tabulated that provides allowed time to get
and return all maintenance tools to handle the job. Similarly, tabulated data
for acquiring the necessary materials to handle the maintenance assignment
should be filed. Machine or facility constant standard data (which is filed by
facility) would appear as shown in Table 4.3.

Table 4.3 Elemental Standard Data for Hand Turret Lathes

No.	Element	Description	Normal time (min)
1.	Loosen and remove extension holder from turret	Starts with reach to wrench, grasp, move to set screw, loosen screw, reach to holder, grasp, move holder aside, release, move wrench aside. Ends with release of wrench.	0.18
2.	Loosen and remove tool or guides from box mill	Starts with reach to socket-head wrench, grasp, move to screw, position, move wrench aside, reach to tool or guides, grasp, move aside. Ends with release of tool or guides.	0.14
3.	Loosen screw chuck	Starts with reach to hammer or wrench, grasp, move to chuck, loosen, lay aside hammer or wrench. Ends with release of hammer or wrench.	0.24
4.	Loosen cross slide	Starts with reach to wrench, grasp, move to study, loosen two, remove front two, move wrench aside. Ends with release of wrench.	1.45
5.	Remove cross slide	Starts with reach to cross slide, grasp, lift to clear, move aside. Ends with release of cross post.	0.23

Variable standard data applied to a facility or machine should be filed independently. For example, Table 4.4 illustrates standard data where the time per foot of weld is dependent on fillet size, electrode size, and the position of the work. Referring to this table, the arc time for a 0.25-in.fillet using a 0.232-in. electrode in flat (F) position is 1.5 min/ft.

PLOTTING CURVES

Because of space limitations, it is not always convenient to tabularize values for variable elements. By plotting a curve or system of curves in the form

Table 4.4 The Welding Institute Standard Data (Fillet Weld Manual Metal Arc; Class E3–Rutile Coated Electrodes–All Positions; Electrode Efficiency–94%)

Fillet size leg length (in.)	Position symbol	No. of runs	Procedure (Page 3P)	Run no.	Awg	Electrode size (in.)	Typical current (ac)	Feet of electrode per foot of fillet (2-in. stub length)	Excess weld metal (%)	Weight of weld metal per foot of fillet (lb)	No. of electrodes per foot of fillet	Typical de-slag and brush time per foot of fillet (min)	Typical arc time (min per foot of fillet)
0.20	H	1	A	1	8	0.160	190	1.8	47	.099	1.2	0.6	1.7
0.20	H	1	A	1	6	0.192	220	1.2	47	.099	0.8	0.4	1.4
0.20	F	1	E	1	6	0.192	220	1.0	17	.079	0.6	0.3	1.2
0.20	V	1	J	1	8	0.160	135	1.8	47	.099	1.2	0.6	2.3
0.20	O	1	P	1	8	0.160	175	1.8	47	.099	1.2	0.6	1.8
0.20	O	1	P	1	6	0.192	200	1.2	47	.099	0.8	0.4	1.6
0.25	H	1	A	1	6	0.192	220	1.8	36	.142	1.2	0.6	2.0

0.25	F	1	E	1	4	0.232	260	1.0	16	.121	0.7	0.4	1.5
0.25	F	1	E	1	—	0.250	290	0.9	16	.121	0.6	0.3	1.3
0.25	V	1	J	1	8	0.160	135	2.5	36	.142	1.7	0.9	3.4
0.25	O	2	Q	1	6	0.192	200	1.2			0.8		
0.375	H	3	B	1 to 3	8	0.160	175	0.8	36	.142	0.5	0.7	2.4
0.375	H	3	B	1 to 3	6	0.192	220	3.7	26	.296	2.5	1.3	4.4
0.375	H	3	B	1 to 3	4	0.232	260	2.5	26	.296	1.6	0.8	3.6
0.375	F	1	E	1	—	0.250	270	2.2	26	.296	1.5	0.8	3.3
0.375	F	1	E	1	4	0.232	260	2.3	13	.266	1.5	0.8	3.3
0.375	F	1	E	1	—	0.250	290	2.0	13	.266	1.3	0.7	2.9
0.375	V	2	K	1 and 2	8	0.160	135	5.2	26	.296	3.4	1.7	7.1
0.375	O	3	R	1 to 3	6	0.192	200	3.6	26	.296	2.4	1.2	4.8

All values expressed in basic minutes.

Weld	Class	Issue	Page
F	E3 –	2	2/1

Source: The Welding Institute.

of an alignment chart, the analyst can express considerable standard data graphically on one page.

The accuracy of the plotted data, the fitting of the best curve, and the expressing of the plotted curve mathematically will have an impact on the reliability of the resulting standard data. The analyst should endeavor to express all curves mathematically for the following reasons.

1.   An algebraic expression eliminates the need to read off a curve, which in some cases is not only difficult but lacks accuracy.
2.   It provides consistency in those cases where a family of curves is plotted.
3.   In cases where two or more curves are dependent on the same controlling variable, the algebraic expression provides the opportunity to simplify the expression by adding the curves algebraically.

Some important points that the analyst should observe when plotting curves include

1.   It is standard procedure to plot time on the ordinate of the charting paper, and the independent variable on the abscissa.
2.   All scales should begin at zero to present their true proportions and to prevent the error of a negative time intercept.
3.   Do not plot curves which, when extended, would show negative time values.
4.   Get observations at extremes of curve in order to be sure of the proper shape of the curve.
5.   When plotting the data, do not extend the curve beyond the range of observations.
6.   In most graphs involving time, the variable element will contain some handling time by the worker. Therefore an intercept greater than zero will exist since this handling time usually is relatively constant.
7.   In some time curves where it has been possible to remove the constant manual time, the curve will go through the origin.
8.   There are many situations which can be represented by a family of curves instead of several separate curves (see Fig. 4.2).
9.   In those cases where a family of curves exist that make up an operation, it is possible to add the curves algebraically (see Fig. 4.3).
10.   Curves plotted on the same x and y axes that intersect do not represent a true family of curves.

Figures 4.4 through 4.9 illustrate the six principal types of curves. An example of how an analyst can use a curve and an algebraic expression to

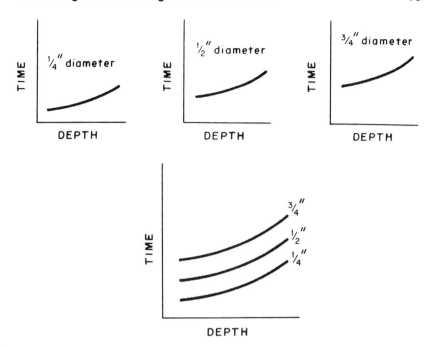

Figure 4.2 Family of curves for drilling various depths using ¼-in., ½-in., and ¾-in. drills.

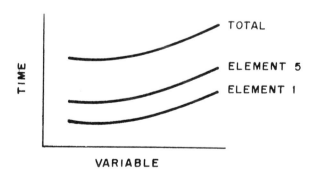

Figure 4.3 A family of curves may be added algebraically.

Type I Linear form, y = mx + b

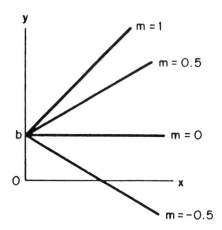

Figure 4.4 Linear form with intercept b and slopes of 1, 0.5, 0, and -0.5.

Type II Power form, $y = bx^m$

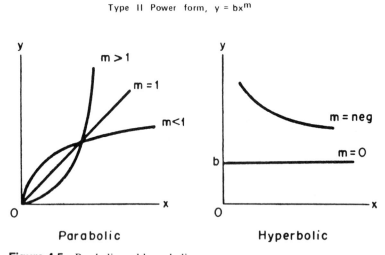

Figure 4.5 Parabolic and hyperbolic curves.

Type III Modified power form, $y = bx^m + c$

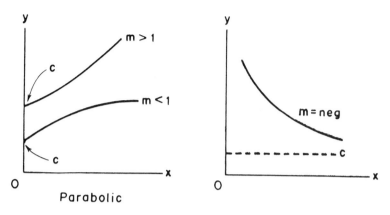

Parabolic

Figure 4.6 Modified parabolic form with positive and negative slopes.

Type IV Exponential form, $y = bm^x$

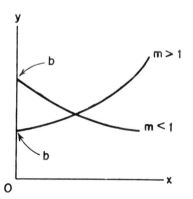

Figure 4.7 Exponential form with slopes less and greater than one.

Type V Modified exponential, $y = bm^x + c$

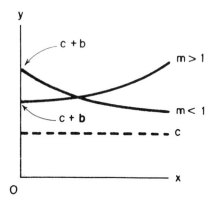

Figure 4.8 Modified exponential relationship with slopes greater and less than one.

Type VI Second degree polynomial, $y = ax^2 + bx + c$

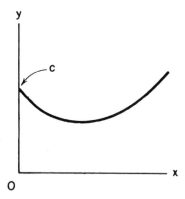

Figure 4.9 Second-degree polynomial.

estimate maintenance work rapidly follows. A maintenance planner decided to develop a method for estimating fillet welding time in the flat position for various fillet sizes and lengths of weld using coated electrodes. He studied five different jobs where the size of the weld varied from fillet leg lengths of 0.125 to 1.0 in. He tabulated the data from these five studies as shown.

Study no.	Fillet leg length (in.)	Arc time (min/in.)
1	0.125	0.12
2	0.250	0.15
3	0.500	0.37
4	0.750	0.93
5	1.000	1.52

Next, he plotted the size of weld against minutes per inch of weld on semilogarithmic paper (see Fig. 4.10). It is apparent with this plotting that the planner would be able to estimate the time for any size of fillet weld per inch. By multiplying the value in minutes per inch by the number of inches to be welded, he would have a reliable estimate of the time for the welding.

For the reasons cited earlier, an algebraic expression is preferred for estimating, rather than using the curve (straight line when plotted on semilogarithmic paper). An analysis of this data follows.

The data plotted on semilogarithmic paper takes the form of a straight line with the equation $Y = AB^x$.

Select two points on the plotting:

Point 1: $x_1 = 1$ $\qquad$ $y_1 = \log 1.52$
Point 2: $x_2 = 0.125$ $\qquad$ $y_2 = \log 0.12$

Solve for slope m:

$$m = \frac{y_1 - y_2}{x_1 - x_2} = \frac{\log 1.52 - \log 0.12}{1 - 0.125} = 1.2601$$

Determine the equation of the line:

$$y - y_1 = m(x - x_1)$$
$$\log y - \log 1.52 = m(x - 1)$$
$$\log y - 0.1818 = m(x - 1)$$
$$\log y = 1.2601x - 1.0783$$

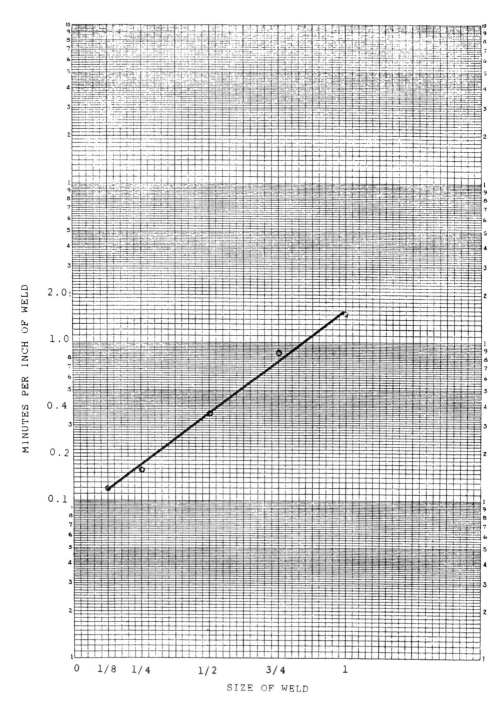

Figure 4.10 Welding time per inch of weld for five radius sizes of weld.

And from the exponential form, $Y = AB^x$,

$$\log y = \log A + x \log B$$

Therefore

$$\log A = -1.0783 \quad \text{and} \quad A = 0.0835$$
$$\log B = 1.2601 \quad \text{and} \quad B = 18.2012$$

Our equation then is: $Y = (0.0835)(18.201^x)$. Testing this equation for a 0.5-in. fillet weld:

Time for inch of weld $= (0.835)(18.201^{0.5}) = 0.3562$ min

This checks quite closely with our graph (Fig. 4.10). To estimate the arc time for an 18-in. length of fillet:

$$(18)(0.3562) = 6.412 \text{ min}$$

METHODS-TIME MEASUREMENT

Fundamental motion times have been successfully used by the practicing industrial engineer to establish standards for direct labor operations since 1945. These fundamental motion times are a collection of valid time standards assigned to fundamental motions and groups of motions that cannot be precisely evaluated with stopwatch time study procedures. They are the result of studying a large sample of diversified operations with a timing device such as a motion picture or video camera that is capable of measuring very short elements.

Several fundamental motion time systems have been broadened to apply to the economic study of long cycle activities such as maintenance and other indirect work. Notable among these systems is Work-Factor where the "Ready" Work-Factor system and the "Brief" Work-Factor system have particular application in establishing standards for maintenance operations. Also, the Methods-Time Measurement system has been broadened so that its MTM-2 and MTM-3 systems have considerable application in maintenance work.

MTM-2, defined by the MTM association of the United Kingdom, is a system of synthesized MTM data and is the second general level of MTM data. It is based exclusively on MTM and consists of

1. Single basic MTM motions
2. Combinations of basic MTM motions

The data is adapted to the operator and is independent of the workplace or equipment used. It is not possible to replace any element in MTM-2 by means of other elements in MTM-2. MTM-2 is recommended for application where

1. The effort portion of the work cycle is more than one minute in length (thus almost all maintenance work is applicable).
2. The cycle is not highly repetitive.
3. The manual portion of the work cycle does not involve a large number of either complex or simultaneous hand motions.

MTM-2 recognizes ten classes of actions, which are referred to as *categories*. These ten categories and their symbols are as follows.

Category	Symbol
Get	G
Put	P
Get Weight	GW
Put Weight	PW
Regrasp	R
Apply Pressure	A
Eye Action	E
Foot Action	F
Step	S
Bend and Arise	B
Crank	C

When using MTM-2, distances are estimated by classes and affect the time of the Get and Put categories. The distances moved are based on the length of the path traveled by the knuckle at the base of the index finger in the case of hand motions or measured at the finger tip if only the fingers move.

The codes for the five tabulated distance classes are as follows.

In.	Code
0–2	5
Over 2–6	15
Over 6–12	30
Over 12–18	45
Over 18	80

The categories Get and Put are usually considered simultaneously. Three variables effect the time required to perform both of these variables. These variables are the case involved, the distance traveled, and the weight handled. The reader should recognize that Get can be considered a composite of reach, grasp, and release while Put is a combination of move and position.

Three cases of Get have been identified as A, B, and C. Case A implies a simple contact grasp such as when the fingers push a screwdriver across a work table. If an object such as a pair of pliers is picked up by simply enclosing the fingers with a single movement, a case B grasp is employed. If the type of grasp is neither an A nor a B, then a case C Get is being employed.

The analyst can resort to a decision diagram (Fig. 4.11) to assist in the determination of the correct case of Get.

Tabular values in TMUs (one TMU = 0.00001 hr) of the three cases of Get applied to each of the five coded distances are shown in Table 4.5.

Just as there are three cases of Get, there are three cases of Put. The decision diagram in Fig. 4.12 may be utilized to determine which case of Put is applicable. An explanation of the three cases of Put, as well as tabular values for each class applied to the five coded distances, is given in Table 4.6.

Put is accomplished in one of two ways: insertion and alignment. An insertion involves placing one object into another, such as a shaft into a sleeve, while an alignment involves orienting a part on a surface such as bringing a straight edge up to a line.

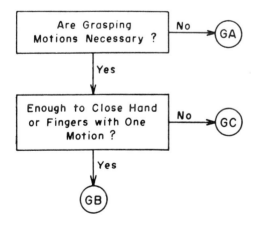

Figure 4.11 Decision diagram to assist in the determination of the correct case of Get.

Table 4.5 Tabular Values for Get

		Distance code				
Case	Description	5	15	30	45	80
GA	No grasp necessary	3	6	9	13	17
GB	Simple closing of fingers or hand	7	10	14	18	23
GC	Any other grasp	14	19	23	27	32

Values in TMUs (1 TMU = 0.00001 hr).

Tolerance limits have been established to provide guidelines as to the most appropriate case of Put. When one part is inserted into another and more than a 0.4-in. clearance exists, the parts can usually be brought together by utilizing a smooth, continuous motion, and consequently a case PA would be involved. If the clearance were reduced to less than 0.4 in. so that some irregularity in motion pattern occurred, yet no pressure was required to mate the objects, the case would be PB. And if a close fit were involved where slight pressure was required to mate the objects or if a loose fit and a difficult to handle part were encountered, thus necessitating corrective mating, the case would be cataloged as PC.

For alignment type Put, the following dimensional relations can be used as guidelines. If an object is placed adjacent to a line to a tolerance less than 1/16 in., the case is classified as PC. However, if the tolerance is more than 1/16 in. but less than 1/4 in., the case is classified as PB, and if the tolerance is more than 1/4 in., the case is classified as PA.

Weight is considered in MTM-2. The time value addition for Get Weight (GW) has been estimated as 1 TMU per effective kilogram. Thus, if a load

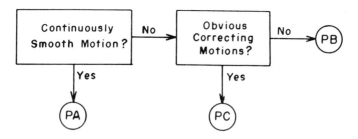

Figure 4.12 Decision diagram to assist in the determination of the correct case of Put.

Table 4.6 Tabular Values for Put

		Distance code				
Case	Description	5	15	30	45	80
PA	Smooth continuous motion	3	6	11	15	20
PB	Some irregularity in motion pattern	10	15	19	24	30
PC	Correction motions obvious	21	26	30	36	41

Values in TMUs (1 TMU = 0.00001 hr).

of 6 kg is handled by both hands, the time addition due to weight would be 3 TMUs since the effective weight per hand is 3 kg.

For Put Weight (PW), additions have been estimated at 1 TMU per 5 kg of effective weight up to a maximum of 20 kg.

The category Regrasp (R) has been defined as opening and closing the fingers to secure better control. Here, a time of 6 TMUs has been assigned. The authors of MTM-2 point out that for a regrasp to be in effect, control must be retained by the hand.

Apply Pressure (A) has been assigned a time of 14 TMUs. The authors point out that this category can be applied to any member of the body and that the maximum permissable movement for an apply pressure is one-quarter in.

Eye Action (E) is allowed under either of the following cases.

1. When it is necessary for the eye to move in order to see the various aspects of the operation involving more than one specific section of the work area
2. When the eye must concentrate on an object so as to recognize a distinguishable characteristic

The estimated value of E is 7 TMUs. The value is only allowed when E must be performed independently of hand or body motions.

Crank (C) occurs when the hands or fingers are used to move an object in a circular path of more than a one-half revolution. A Put is indicated in cranks of less than a one-half revolution. Only two variables are associated with the category crank under MTM-2. These are the number of revolutions and the weight or resistance involved. A time of 15 TMUs is allotted for each complete revolution. Where weight or resistance is significant, PW is applied to each revolution.

Foot movements are allowed 9 TMUs, and step movements 18 TMUs. The time for step movement is based upon a 34-in. pace. The decision diagram in

Fig. 4.13 can be helpful in ascertaining whether a given movement should be classified as a Step (S) or a Foot Action (F).

The category Bend and Arise (B) occurs when the body changes its vertical position. Typical movements characteristic of B include sitting down, standing up, and kneeling. A time value of 61 TMUs has been assigned to B. The authors indicate that when an operator kneels on both knees, the movement should be classed as 2 B.

A summary of MTM-2 values is shown in Table 4.7. The reader should recognize that motions performed simultaneously with both hands cannot always be performed in the same time as motions performed by one hand only. Figure 4.14 reflects motion patterns where the time required for simultaneous motions is the same as that required for motions performed by one hand. In these instances an X appears. When additional time is required to perform simultaneous motions, the magnitude of this extra amount is given with the appropriate coding.

As with all fundamental motion data systems, the novice should not try to apply the data until being properly trained in its use and application.

MAYNARD OPERATION SEQUENCE TECHNIQUE

An outgrowth of MTM referred to as MOST (Maynard Operation Sequence Technique) is a simplified system developed and originally applied at Saab-Scania in Sweden in 1967 by Kjell B. Zandin. H. B. Maynard & Co., which is currently marketing MOST, states that standards can be established at least five times faster when using MOST as compared to using MTM-1, with little if any sacrifice in accuracy. Thus MOST competes with MTM-2 as a technique for the economical establishment of standards on indirect labor such as mainte-

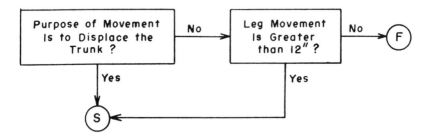

Figure 4.13 Decision diagram to assist in the determination of Step or Foot Action.

Table 4.7 A Summary of MTM-2 Data

Code	Distance code				
	5	15	30	45	80
GA	3	6	9	13	17
GB	7	10	14	18	23
GC	14	19	23	27	32
PA	3	6	11	15	20
PB	10	15	19	24	30
PC	21	26	30	36	41
GW	(1 TMU per effective kg)				
PW	(1 TMU per 5 kg)				
A	(14 TMUs per occurrence)				
R	(6 TMUs per occurrence)				
E	(7 TMUs per occurrence)				
C	(15 TMUs per occurrence)				
S	(18 TMUs per occurrence)				
F	(9 TMUs per occurrence)				
B	(61 TMUs per occurrence)				

Values in TMUs (1 TMU = 0.00001 hr).

nance. However, MOST utilizes larger blocks of fundamental motions than MTM-2, and consequently, analysis of the work content of an operation can be made faster. MTM-2 is built around thirty-seven time values (see Table 4.7) for describing manual work, but MOST utilizes only sixteen time fragments. MOST identifies three basic sequence models: General Move, Controlled Move, and Tool Use.

The General Move sequence is used for identifying the special free movement of an object through the air, while the Controlled Move sequence describes the movement of an object when it remains in contact with a surface or is attached to another object during the movement. The Tool Use sequence has been developed for the use of common hand tools.

In order to identify the exact way a General Move is performed, four subactivities are considered: action distance, which is primarily a horizontal distance; body motion, which is mainly vertical; gain control; and place.

Index numbers that are time related are assigned to the applicable subactivity. MOST uses the index numbers of 0, 1, 3, 6, 10, and 16. The reader can appreciate that it is relatively easy to memorize these values and their application to the four subactivities of General Move.

MTM-2 Simultaneous
Hand Motion Allowances

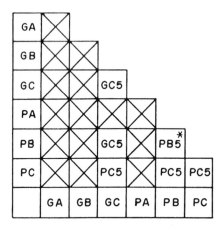

* If PB __ is performed
simultaneously with PB __,
an addition of PB5 is made
only if the actions are
outside the normal area
of vision.

Figure 4.14 Additional times to be added with hand motion patterns when utilizing both hands. When X is shown, no additional time is required.

It has been estimated that about 50% of manual work occurs as General Move. The activities of walking to a location, bending to pick up an object, arising after bending, placing an object, reaching, and gaining control of an object are all illustrations of General Move.

The Controlled Move sequence is used in conjunction with such manual operations as cranking, pulling a starting lever, turning a steering wheel, engaging a starting switch, etc. In performing Controlled Move sequences, the following subactivities may pertain: action distance, body motion, gain control, move controlled, process time, and alignment.

The final sequence used in MOST is the Tool/Equipment Use model. Work using tools, such as cutting, fastening, welding, gaging, and writing, is covered by this sequence. The Tool/Equipment Use model engages a combination of General Move and Controlled Move activities. Other subactivities unique to this activity include fasten, loosen, surface treat, record, think, and measure.

The trained maintenance analyst can successfully use any of the fundamental motion data techniques (MOST, MTM-2, Work-Factor, etc.) to establish reliable maintenance standards on all planned effort elements. Those process-controlled elements can be estimated accurately from knowledge of equipment feeds and speeds. Also, it should be recognized that it is not necessary to plan each maintenance work order in minute detail representative of the fundamental motion elements. Broader element standards can be computed from the fundamental motion data and stored in the standard data bank for use in estimating broader elements listed on the work order. For example, the element "walk approximately 40 ft to tool crib and acquire standard tool kit, return 40 ft to machine center" can have an accurate standard based upon any of the standard data techniques. This value can then be used when the planner develops a standard for a repair operation on a given machine. He will not need to break this element down into the many fundamental motions that make up this effort.

UNIVERSAL MAINTENANCE STANDARDS BASED UPON SLOTTING

Although reliable standards can be established on maintenance work in advance of performing the work through the use of standard data, the cost of the development of the standards may be prohibitive. An alternative that has proven satisfactory in many cases that is considerably less costly to apply is through the use of universal maintenance standards based on slotting. The principle behind universal maintenance standards is assigning the majority (perhaps as much as 90%) of the maintenance operations to appropriate groups (slots). Each group will have its own standard, which usually is the mean time for all maintenance operations that have previously been assigned to the particular group. The standard time for each standard (benchmark) in the group was determined by either standard data, fundamental motion data, stopwatch time study, or a combination of any of the measurement techniques. For example, a given group, which we can designate group C, may include the following maintenance operations that have been studied using one of the measurement techniques: "Replace defective 1.5-in. gate valve," and "replace two 8-ft sections of 1-in. galvanized pipe." The standard time computed for any maintenance work order given a group C assignment might be 0.82 hr. This time will represent the mean ($\bar{x}$) of all jobs within the group, and the dispersion of the jobs within the group for plus-or-minus two standard deviations will be some percentage of $\bar{x}$ (perhaps $\pm$ 10%). Thus if a maintenance job was assigned in advance of the work performed to the correct group, the time assignment would always be correct within $\pm$ 10%. This accuracy is usually considered quite satisfactory in connection with the development of standards for indirect work.

The three principal considerations in establishing a universal maintenance standard system are

1. Determination of a number of standards (groups or slots) to do a satisfactory job. (Approximately twenty slots should be used in establishing maintenance standards where the time to complete the order takes up to 40 hr.)
2. Determination of the numerical standards that will be representative of each group of operations contained in each slot.
3. The correct assignment of each maintenance work order as it occurs to the appropriate slot.

The first step is to determine valid benchmark standards based upon measurement of an adequate sample of maintenance work for which the universal maintenance standards system is being developed. This is the most time-consuming and costly step in the installation of a universal maintenance standards program. To do this, it will be necessary to establish a relatively large number of standards (200 or more) that are representative of the entire population of maintenance work. These measured standards, often referred to as benchmark standards, should be carefully developed by competent analysts using proven industrial engineering tools, including stopwatch time study, standard data, formulas, and fundamental motion data.

Once established, the maintenance benchmark standards are arranged in numerical sequence. Thus, assuming there were 200 benchmark standards, the shortest would be listed first, the next shortest second, and so on until the longest would be listed last. If 20 slots were used and the distribution is based upon the standard normal variable, there would be 20 equal intervals (truncation of the two tails allow this). For example, the standard normal variable may have a truncated range of: $-3.0 \leq z \leq +3.0$ which accounts for 99.87% of the area under the curve (see Table 4.8). The range of each interval would be 6 sigma/20 = 0.3 sigma. The number of benchmark standards used in the compilation of the mean of each of the 20 slots (intervals) would equal p (z E interval)(200)/0.9987. Thus slot number ten and eleven (because of symmetry) would have

$p(-0.3 \leq z \leq 0.0)(200)/0.9987$

$= |\ 0.3821 - 0.5000\ |\ (200)/0.9987 = 23.61$ standards

Fractions will be rounded off so that all 200 benchmark standards are assigned to a slot. The universal maintenance standard time for each slot is calculated as the mean of the benchmark standards assigned to the slot.

Table 4.8 Areas of the Normal Curve

z	Area	z	Area
− 3.0	.0013	0.1	.5398
− 2.9	.0019	0.2	.5793
− 2.8	.0026	0.3	.6179
− 2.7	.0035	0.4	.6554
− 2.6	.0047	0.5	.6915
− 2.5	.0062	0.6	.7257
− 2.4	.0082	0.7	.7580
− 2.3	.0107	0.8	.7881
− 2.2	.0139	0.9	.8159
− 2.1	.0179	1.0	.8413
− 2.0	.0228	1.1	.8643
− 1.9	.0287	1.2	.8849
− 1.8	.0359	1.3	.9032
− 1.7	.0446	1.4	.9192
− 1.6	.0548	1.5	.9332
− 1.5	.0668	1.6	.9452
− 1.4	.0808	1.7	.9554
− 1.3	.0968	1.8	.9641
− 1.2	.1151	1.9	.9713
− 1.1	.1357	2.0	.9772
− 1.0	.1587	2.1	.9821
− 0.9	.1841	2.2	.9861
− 0.8	.2119	2.3	.9893
− 0.7	.2420	2.4	.9918
− 0.6	.2741	2.5	.9938
− 0.5	.3085	2.6	.9953
− 0.4	.3446	2.7	.9965
− 0.3	.3821	2.8	.9974
− 0.2	.4207	2.9	.9981
− 0.1	.4602	3.0	.9987
0.0	.5000		

When planning a new maintenance work order, the analyst will be able to fit the job to a slot where similar jobs have been studied and standards established.

Those maintenance jobs making up the benchmark standards should be carefully defined so that the analyst doing the planning will have little difficulty in selecting the correct slot for the work order under study. In describing the work content of the benchmark jobs, information should be provided on the three divisions of effort that comprise every maintenance work order. These are

1. Transportation
2. Direct work
3. Indirect work

Transportation refers to work performed in movements by the maintenance worker between the shop and the job from job to job, and from job back to the maintenance shop. Here the travel can be horizontal, vertical, or a combination of horizontal and vertical. Typical transportation work elements include such activity as "walk up and down two flights of stairs," "push truck loaded with tools five hundred feet," "ride on motor truck half mile," "ride elevator to 12th floor," etc. The amount of transportation required and the standard time allowed for this transportation should be clearly shown in the description of the benchmark job.

Direct work represents those elements of work that advances the repair, overhaul, or maintenance operation. These work elements are easily discernible in the performance of the job. For example, in the installation of a new door by a maintenance carpenter, the direct work elements may include "cut door to rough size," "plane length and width to finish size," "locate three hinge areas and mark," "chisel out hinge areas," "mark for screws (6 for each hinge)," "install hinges on door," "mark for lock," "drill for lock," "install lock," etc.

Indirect work represents those work elements that must be performed in order to complete the maintenance job but as a general rule cannot be evaluated by physical evidence in the completed job or at any stage during the course of the work, except by deductive inference as implied by certain characteristic features of the job. Indirect work elements will fall into these classifications:

1. Tooling
2. Material
3. Planning

Typical tooling elements that may occur include "get and check equipment," "clean tools," "repair tools," "return equipment at end of job."

Materials elements may well include such items of work as "get and check materials," "pick up and dispose of scrap," "make minor repairs to materials."

Planning represents those elements where the maintenance worker is obliged to consult or make some analysis or test before he can proceed with additional direct work. Typical planning elements include "consult with foreman," "inspect, check, and test," "plan procedures of work."

The reader can understand that two work orders, each for the same work but performed in different areas may well require different standards in view of the difference in transportation. Without complete information on the benchmark standards and complete information on the new work order being planned, the slotting technique will not succeed.

A plotting of the benchmark standards (frequency versus time) will always reveal a skewed distribution toward an increase in time. It would seem that a skewed distribution may outperform the normal in the establishment of slot values. The gamma distribution is a positively skewed distribution with a probability density function given by

$$f(x) = \frac{1}{0\ \beta^2\Gamma(\alpha)}x^{\alpha - 1}\ e^{-x\beta} \qquad \text{for } x, \alpha, \beta > 0$$

elsewhere, where $\Gamma(\alpha)$ is a value of the gamma function given by

$$\Gamma(\alpha) = \int_0^\infty x^{\alpha - 1}\ e^{-x}\ dx$$
$$= (\alpha - 1)\Gamma(\alpha - 1)$$

The skewness of the gamma distribution decreases as α increases for a fixed β.

The mean and the variance of this distribution can be shown to equal:

$$\mu = \alpha\beta$$
$$\sigma^2 = \alpha\beta^2$$

For a given set of data, estimates of α and β are determined by first obtaining the mean and variance of the data. The estimates can then be calculated as

$$\hat{\alpha} = \frac{\mu^2}{\sigma^2}$$

$$\hat{\beta} = \frac{\mu}{\hat{\alpha}}$$

An alternative to using the gamma distribution is to plot the distribution of the benchmark standards (frequency versus time) and then divide this plotting into the desired number of slots (perhaps 20). The proportional area of each slot to the total area will equal that proportion of the total benchmark standards that will be used to compute the time value of that slot.

The application of universal indirect standards offers an opportunity to introduce standards for a majority of maintenance operations at a moderate cost and also minimizes the cost of maintaining the maintenance standards system.

SUMMARY

Standards based on stopwatch studies, standard data, fundamental motion data, and slotting will provide the time it should take to do a maintenance job under average conditions with a qualified workman using an appropriate method and experiencing normal delays under local operating conditions.

It is essential that reliable standards be estimated when the maintenance work is planned and that these values be shown on the work order. These estimates should be based upon measurement methods and should not be provided on the basis of the judgment alone of the maintenance planner. Thus tabulated data that has been developed from stopwatch time studies or fundamental motion data should be used for the development of a fair standard for each operation taking place on the maintenance job. A computer should be utilized to handle the arithmetic operations in order to save time and minimize errors.

A second alternative for the rapid development of good standards during the planning process for release on the work order is the slotting technique. Here the success of the installation is dependent primarily on the accuracy of the benchmark standards used in the development of the distribution (frequency versus time). These benchmark standards can be developed from stopwatch time studies and fundamental motion data.

Standards are essential for the effective operation of a maintenance department. They permit the accurate measurement of the work load on each craft and thus help maintain a steady work force and help maintain a steady level of good maintenance service to the production areas of the plant. Good standards are essential for scheduling jobs, the evaluation of suggestions and improvements, measuring the performance of each craft, locating delays, and developing improvements. And they are essential in connection with sound wage incentive systems.

SELECTED BIBLIOGRAPHY

Antis, William, Honeycutt, John M. Jr., and Koch, Edward N. *The Basic Motions of MTM*, 3rd ed. Pittsburgh: The Maynard Foundation, 1971.

Barnes, Ralph M. *Motion and Time Study: Design and Measurement of Work*, 7th ed. New York: Wiley, 1980.

Clerk, Forest D., and Lorenzoni, A. B. *Applied Cost Engineering*, 2nd ed. New York: Dekker, 1985.

Karger, Delmar W., and Bayha, Franklin H. *Engineered Work Measurement*. New York: Industrial Press, 1957.

Mundel, Marvin E. *Motion and Time Study: Improving Productivity*, 5th ed. Englewood Cliffs, NJ: Prentice-Hall, 1978.

Niebel, B. W. *Motion and Time Study*, 9th ed. Homewood, IL: Richard D. Irwin, Inc., 1993.

Zandin, Kjell B. *MOST Work Measurement Systems*. New York: Dekker, 1980.

5
Estimating Materials Costs in Maintenance Work

The planning function in maintenance work identifies not only what is required, but how it should be accomplished. The former involves identification of the problem, its location, and its priority. The latter includes the specification of the methods to be used and the manning requirements. Methods to be used include the identification of the material and tool requirements as well as standard times in connection with the work. In the previous chapter, the development of standard times was discussed. In this chapter, we review procedures in identifying the materials and their cost for a specific maintenance work order. Careful planning in the selection and control of maintenance materials presents an important consideration in a sound maintenance engineering management program. Typically, maintenance materials represent 40–50% of total maintenance cost in U.S. industry.

We have learned that the work order should be as brief as possible; it should provide only those details for adequately making the repair and for accurate costing. It should provide for both an estimate of labor requirements and materials needed. Thus with good material planning in the utilization of both stock (repair materials) and stores (spare parts), cost improvement will be an ongoing process, rush repair jobs will be expedited, and scheduled maintenance will be completed in accordance with plans.

Stock- and storerooms that are well planned and operated efficiently will result in significant savings in both materials costs and maintenance labor. Work orders will be better scheduled and delays resulting from unavailability of the correct materials and spare parts will be minimal. Cost of excessive inventory and obsolete components is significant in most maintenance storerooms and stockrooms. The reader should understand that the costs for carrying inventory in stock or stores (referred to as possession costs) typically range from 10 to 25% per year.

Store- and stockrooms should provide for convenient location and rapid safe removal of all items. They should be designed to protect all items from damage, theft, and deterioration. In describing the items on store- and stockroom records, mnemonic codes should be used so that items are clearly identified in connection with group technology principles. Each item is described so that anyone familiar with the plant and its facilities will understand what the part is used for, the logical subgroups of which it is a part, and its location in stores.

SELECTING AND COSTING NONSTOCK (STORES) ITEMS

Identification of nonstock items needed to complete the repair or overhaul of equipment being maintained is usually apparent from the work order request and drawings of the equipment. These items include such components as bearings, bushings, bolts, valves, unions, couplings, motors. Usually these items are inventoried in the storeroom and a balance-of-stores ledger will provide information as to their most recent purchase cost. However, this purchase cost does not represent the actual cost, which is the figure needed in pricing the total cost for performing the maintenance work.

The planning function includes identification of all materials needed to perform the repair, the amounts of materials needed, and their availability. The costing of the specified materials will usually be performed in the cost department where the closed work order will be costed for record purposes.

In selecting the stores items to be used for the repair, the planner should always be alert to utilize less expensive components that still meet service requirements. For example, in the repair of a certain line of distribution transformers, it was found that a fuller board plate could be economically substituted for porcelain plates to separate and hold the wire leads coming out of the transformers. The fuller board plate stood up just as well in service yet was considerably less expensive. The planner should keep in mind that standard items such as valves, relays, air cylinders, transformers, pipe fittings, bearings, couplings, chains, hinges, hardware, and motors can often be substituted for specialty components that cost far more. Often a minor adaptation of the equipment will permit the use of standard items.

The planner should always be alert to the possibility of standardizing stores. An effort should be made to minimize the sizes, lengths, shapes, grades, and so on, of all stores. Frequently standard components can replace specialty components at savings. Usually a standard bolt that is a half-inch longer than a specialty bolt of the same diameter is much less expensive. The following typical economies result from reductions in the sizes and grades of stores employed: purchase orders will be for larger amounts which are almost always less expensive; inventories will be smaller since less material must be maintained as a reserve; fewer entries will be made in the stores' records; fewer invoices will need to be paid; fewer spaces will be needed to house materials in the storeroom; sampling inspections will reduce the total number of parts inspected; and fewer price quotations and purchase orders will be prepared.

In costing the various store items used in the repair work, consideration should be given to

1. The latest purchase or make cost.
2. The cost of inventorying the material.
3. The cost of invested capital.
4. The increase in value of the stock items because of inflation.
5. The decrease in value of the stock items because of decay or spoilage. For example, such items as bonded rubber vibration isolators have a limited shelf life.

The latest purchase order or make cost, as has been mentioned, is usually available on the inventory card or balance-of-stores ledger card. If this information is not recorded here, it can readily be obtained from the purchasing department.

The cost of inventorying the material is a cost that usually is not accurately known by even the most advanced industries. However, it is an important and significant cost that should be estimated and be applied to all stock and stores. Typically, inventorying cost is computed on the basis of the amount of floor or cubic space the item requires while in inventory, and the average amount of time the item spends in inventory. For example, a bin occupying 6 ft^2 may contain on an average ten 1.5-in. gate valves. The average time a gate valve spends in inventory is two months. If the inventorying cost of floor space is equal to \$4.00/ft^2/yr, the inventorying cost of one 1.5-in. gate valve would be

$$C_i = \frac{Is}{NM}$$

where

I = cost of floor space per ft^2 per year

s = size of bin (ft^2)

N = average number of items stored in bin

M = reciprocal of years item spends in inventory

C_i = inventory cost per item

Thus

$$C_i = \frac{(\$4.00)(6)}{(10)(6)}$$

$$C_i = \$0.40$$

The cost of invested capital takes into consideration the time value of money. Thus a \$15.00 inventory item whose average time in inventory is three years would have a value at time of withdrawal from inventory of

$$S = C_p(1 + i)^n$$

where

S = worth of inventory items n periods later

C_p = present worth of the inventory item (purchase price plus delivery cost)

n = number of interest periods

i = interest rate for a given period

If interest rates are estimated at 15% annually, then

$$S = 15(1 + 0.15)^3$$

$$S = \$22.81$$

Not all engineering economists hold that first cost should be the basis for interest. Some support the notion of using an average interest, which is the interest on the average book value during the life of the asset. Other models support the use of an "exact method" where a sinking fund depreciation combined with interest on first cost is used.

Since we are looking at interest as expected return, we are using the initial cost and prevailing interest rates for securing capital as the basis of the worth

Table 5.1 Compound Amount Factor, $(1 + i)^n$, to Determine Future Worth of a

Periods	.07	.08	.09	.1	.11	.12
1	1.070000	1.080000	1.090000	1.100000	1.110000	1.120000
2	1.14490	1.16640	1.188100	1.210000	1.232100	1.254400
3	1.225043	1.259712	1.295029	1.331000	1.367631	1.404928
4	1.310796	1.360489	1.411582	1.464100	1.518070	1.573519
5	1.402552	1.469328	1.538624	1.610510	1.685058	1.762342
6	1.500730	1.586874	1.677100	1.771561	1.870415	1.973823
7	1.605781	1.713824	1.828039	1.948717	2.076160	2.210681
8	1.718186	1.850930	1.992563	2.143589	2.304538	2.475963
9	1.838459	1.999005	2.171893	2.357948	2.558037	2.773079
10	1.967151	2.158925	2.367364	2.593742	2.839421	3.105848

of inventory items after n periods. Table 5.1 provides values of $(1 + i)^n$ for typical interest rates and periods up to 10 yr.

The time value of money concept is most useful in determining the amount of inventory (stock and/or stores) to be ordered or procured. For example, the plant engineer may be confronted with the alternative whether to run four ducts at the present to remote sections of the plant or run one duct now, a second five years hence, a third in ten years, and a fourth in fifteen years. These time estimates are the best estimates that have been provided by management's long-range planning activity. All ducts are over a thousand feet in length. Through large lot purchasing and consequent economies of scale based on the manufacturing progress function, it is estimated the cost of installing one duct is $1.82 per ft while the cost of installing the four ducts is $1.15 per ft.

By summing the present worths of all separate investments per foot based upon 10% interest:

Time of investment	Four ducts now	One duct at a time
Now	(4)(1.15) = $4.60	$1.82
In 5 yr	—	($1.82)(0.621) = $1.13
In 10 yr	—	($1.82)(0.386) = $0.70
In 15 yr	—	($1.82)(0.239) = $0.44
Present worth/ft of duct	$4.60	$4.09

Single Amount—For Present Worth, Use Reciprocal Factor

.13	.14	.15	.16	.17	.18
1.130000	1.140000	1.15	1.160000	1.170000	1.180000
1.276900	1.299600	1.322500	1.345600	1.368900	1.392400
1.442897	1.481544	1.502875	1.560896	1.601613	1.643032
1.630474	1.688960	1.749006	1.810639	1.873887	1.938778
1.842435	1.925415	2.011357	2.100342	2.192448	2.287758
2.081952	2.194973	2.313061	2.436396	2.565164	2.699554
2.352605	2.502269	2.660020	2.826220	3.001242	3.185474
2.658444	2.852586	3.059023	3.278415	3.511453	3.758859
3.004042	3.251949	3.517876	3.802961	4.108400	4.435454
3.394567	3.707221	4.045558	4.411435	4.806828	5.233836

It is economically advantageous to install one duct every five years to meet needed expansion requirements rather than put in all four ducts at this time. This decision is also favored by certain intangibles. For example, there is always some risk in connection with the forecasting of future needs. Then too, there is the possibility of unforeseen obsolescence and maintenance.

Another thought to consider when estimating the cost of stores items is an allowance for inflation. Although interest rates usually reflect inflation levels, it is usually wise to consider inflation independent of the prevailing interest rates. The rate of inflation has not been static over the past ten years and probably will continue to be a variable that will need to be reckoned with. Models that allow 1% per month of purchase cost while the item is in inventory today are considered reliable for many items.

Finally, some allowance for shelf life should be considered for spoilage, deterioration, theft, and obsolescence. A 10% of cost allowance is typical for many inventory items.

By combining the aforementioned five considerations of cost applying to stores, we get

$$C_s = C_p + C_i + (S - C_p) + (0.01)(T)(C_p) + 0.1C_p$$

or

$$C_s = S + C_i + \left(\frac{TC_p + 10C_p}{100} \right)$$

where

T = time in months the stock item is in inventory

C_s = total cost of stock or stores at time of repair

SELECTION AND COSTING OF STOCK

Stock that will be needed to complete the repair should be identified during the planning function. Typical stock items include bar stock, tubing, sheet metal, and so on. All materials that must be sized for the repair work under consideration are identified as stock.

It is in the specification of the stock to be used that the planning function can be extremely helpful not only from the standpoint of economics but in identifying a suitable stock that is readily available. For example, it is usually advantageous to substitute a slightly more costly material that is available than to specify a less expensive one that will adequately do the job but will need to be ordered.

When specifying the stock to be used, the planner must take into consideration the work elements involved in processing the material. Although a given material may be available, it may be too thick to readily form to the contour required or it may be too thin and will tear in service. When the planner substitutes any material that results in poorer physical or mechanical properties characteristic of the original installation, he should obtain approval of plant engineering prior to releasing the work order. There will be many opportunities that will make the facility or equipment more reliable after the repair is completed than it was when new. Those repairs and overhauls that improve upon the production capability and reliability of a facility are indicative of sound engineering maintenance management.

The stock material cost is determined by first calculating the present material cost, C_p, and then expanding this value to include the cost of inventorying the material, the cost of the invested capital to produce the stock, and the increase in value of the stock due to inflation.

$$C_p = WP(1 + L_1 + L_2) - M$$

where

W = weight (or other unit of quantity) of material used

P = purchase price of material per unit (this is the delivered price)

L_1 = losses due to scrap, chips, skeletons, etc.

L_2 = losses due to unused stock returned to inventory but too small for future use

M = unit price of salvage material

The cost per unit weight (or other unit of quantity, such as length, volume, etc.) of the material is usually posted on the inventory card or may be obtained from the purchasing department. The amount of material actually going into the repair is estimated from drawings of the equipment being repaired and the description of the repair outlined on the work order.

The amount of losses due to scrap, chips, skeletons, and so forth, include those losses that may be encountered because of spoilage or rejection by quality control due to inferior repair. Also included are the chips from machining, skeletons from stamping, short ends from turning and shearing, and others.

Those losses due to unused stock returned to inventory, but too small for future use, usually represent a significant loss in maintenance work. For example, after sawing a 3.75-ft piece of channel stock there will be a 2.25-ft piece left from a section 6 ft in length. This 2.25-ft piece may have no utility in future maintenance work since the company may only have use for channel stock of size 2.5 ft and greater. In this example, L_2 would be equal to 2.25, assuming the unit of quantity is feet.

The unit price of material salvaged should be deducted from the computation of stores and stock utilized in equipment repair. This is especially true if the salvage material has any significant value such as parts made of brass or bronze. On many repair items, the replaced item has practically no value and in such cases this item may be dropped from the cost equation.

Once C_p is calculated for the purchase cost for the stock to be charged to the work order, C_s may be computed as explained earlier.

MATERIALS CONTROL PROCEDURES

In Chapter 6, formal inventory control procedures are discussed. Suffice to bring out at this time that materials control procedures are just as important as sound estimating in the control of maintenance material costs.

The efficient material planner is always considering possible ways to minimize inventory needs. Often the plant is located near reliable suppliers who will maintain an adequate inventory of certain items, thus allowing the user to stock a much smaller inventory than would otherwise be needed.

What stock items are to be carried as regular repair stock is a consideration that should be carefully planned. Periodic monitoring to review turnover intervals, shrinkage, and stock designations is important in controlling the optimum size of inventory. Not only what stock items but what specific type should be carried is an ongoing problem. For example, there is the decision whether or not to use treated lumber in connection with the overhaul of certain structures. A certain trestle will cost $10,000 if rebuilt of untreated lumber and will have an estimated life of 20 yr. However, if treated lumber is used, the life of the structure is estimated to be 40 yr and the annual maintenance cost will be reduced by an estimated $200.00. The cost of the overhaul using the treated lumber is estimated to be $15,000. Interest rates are averaging 10%. The maintenance analyst in charge of inventory may make the following economic analysis:

	Untreated lumber	Treated lumber
Investment	$10,000	$15,000
Life	20 yr	40 yr
Difference in annual	$200	—
maintenance cost		
Depreciation	$500	$375
Expected return (10%)	$1,000	$1,500
Annual cost plus interest	$1,700	$1,875

In this case, it is recommended that the less costly untreated lumber be inventoried.

REDUCING INVENTORIES OF SPARE PARTS

There is a natural tendency to overstock spare parts to assure that necessary components are available when equipment breaks down. A sound maintenance management engineering program through its cost improvement activities should always recognize potential savings in inventory to be realized through efficient repair rather than replacement.

Today, modern welding technology makes it possible to weld, braze, or solder the vast majority of metals in any combination. The joint produced can be equal to and often is superior than the original part. Large castings, forgings, and roll-formed components that break can often be repaired at the work site at a fraction of the cost to remove and replace the broken component.

The development, in recent years, of special maintenance filler and overlay alloys requiring minimum heat input has permitted the extensive salvaging of metal components that have failed. Thus the combined use of special welding alloys and traditional metal joining techniques enable the cost-effective procedure of rebuilding parts.

In determining spare part inventories, consideration should be given to the alternative of repair as opposed to replacement. Proper and appropriate training of maintenance personnel is essential if such repair programs are to succeed.

SUMMARY

The true costs of materials and spare parts utilized in maintenance work should be carefully estimated and recorded so that total maintenance cost can be identified, compared, and used as the basis of developing better maintenance methods and improved facility reliability. Procurement costs alone are insufficient information for estimating the cost of both stock and stores. Procurement costs must be coupled with the costs of inventorying the material and this sum must be adjusted to allow for the cost of invested capital, the increase in value of materials because of inflation, and the decrease in value of stock and stores because of decay or storage.

Aggressive plant managers regularly modernize their plants by periodic overhaul and investment in new capital equipment. The paramount question on these material procurement expenditures is: will it pay? Asking this question usually involves the consideration of alternatives. In comparing economy of alternatives it is important that the maintenance manager understands the concept of return on investment.

For

i = interest rate per period

n = number of interest periods

(Thus, \$100 n years from now = $\$100/(1 + i)^n$.)

C = cash receipts

D = cash payments

P = present worth at the beginning of n periods

S = lump sum of money at the end of n periods

R = an end-of-period payment or receipt in a uniform series continuing for n periods

then

$$P = \frac{C_0 - D_0}{(1 + i)^0} + \frac{C_1 - D_1}{(1 + i)^1} + \frac{C_2 - D_2}{(1 + i)^2} + \cdots + \frac{C_n - D_n}{(1 + i)^n}$$

$S = P (1 + i)^n$ single payment

$$P = S \frac{1}{(1 + i)^n}$$

$R = S \dfrac{i}{(1 + i)^n - 1}$ uniform series, sinking fund

$R = P \dfrac{i (1 + i)^n}{(1 + i)^n - 1}$ uniform series, capital recovery

SELECTED BIBLIOGRAPHY

D'Ovidio, Gene J., and Behling, Richard L. "Material Requirements Handling." In Gavriel Salvendy, Ed., *Handbook of Industrial Engineering*. New York: Wiley, 1982.

Fabrycky, W. J., and Thuesen, G. J. *Economic Decision Analysis*. Englewood Cliffs, NJ: Prentice-Hall, 1974.

Orlicky, J. *Material Requirements Planning*. New York: McGraw-Hill, 1975.

Ostwald, Philip F. *Manufacturing Cost Estimating Guide*. New York: McGraw-Hill, 1983.

Wasserman, Rene. *How to Save Millions by Reducing Inventories of Spare Parts*. New York: Eutectic-Castolin Institute for the Advancement of Maintenance and Repair Welding Techniques, 1971.

6
Inventory Control of Maintenance Materials

Perhaps the major complaint of maintenance supervision is the unavailability of spare parts and materials at the time they are needed. With the increasing complexity of modern manufacturing equipment, the cost of inventorying spare parts is high. Maintaining an adequate inventory of parts and supplies and providing a procedure for getting these materials to the work site when needed is essential in a modern engineering maintenance management program. Inventory management is charged with the continuing conflict of reducing inventory dollars while avoiding stockouts.

In this chapter stock refers to materials used for repair and maintenance, such as lumber for carpenters, pipe and tubing for plumbers, paint for painters, and sheet metal for sheet metal workers. Stores refer to all spare parts and hardware items, such as motors, rheostats, bolts, and screws.

ABC CLASSIFICATION SYSTEM FOR INVENTORY

In any maintenance inventory control system, parts and materials for routine maintenance should always be immediately available. Parts for major overhaul and other nonroutine maintenance should be controlled in such a manner that capital invested in inventory is working to the best advantage.

The reader should understand that both the carrying (possession) costs and acquisition (ordering) costs are affected by the number of stock and stores turns per year. For example, referring to Fig. 6.1, note that the cost of acquisition (ordering, receiving, delivering to the storeroom) is very little with one turnover a year, but increases steadily as the number of turnovers increases. On the other hand, the cost of possession (interest on money invested, storage cost including insurance, wages of stock clerks, and losses due to obsolescence of stock and stores) decreases as the number of turns per year increases.

The objectives of effective inventory control are

1. To relate stock and stores quantities to demand, thus avoiding both overstocking or understocking
2. To avoid losses due to spoilage, pilferage, and obsolescence
3. To obtain the best turnover rate on all items by considering both the costs of acquisition and possession

To best classify inventory and acquire the control needed in the least costly manner, we can apply Pareto's law. Vilfredo Pareto, a 19th-century Italian

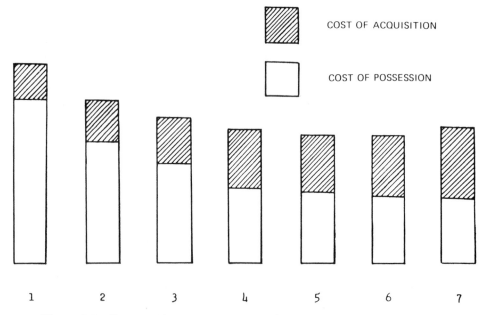

Figure 6.1 Costs resulting from the number of inventory turns per year.

economist articulated the fact that the significant items in a group usually constitute only a small portion of the total number of items in the group. Thus, in the inventory of maintenance stock and stores the major proportion of the total inventory value will usually be comprised of as little as 10% of the items controlled (see Fig. 6.2).

To determine the amount and type of control to establish on all the items inventoried, it is a good idea to classify all maintenance stores and stocks into three categories. These are

Class A. These stocks and stores would represent only between 10 and 15% of the total items yet their monetary value would be between 70 and 85% of the total investment in inventory.

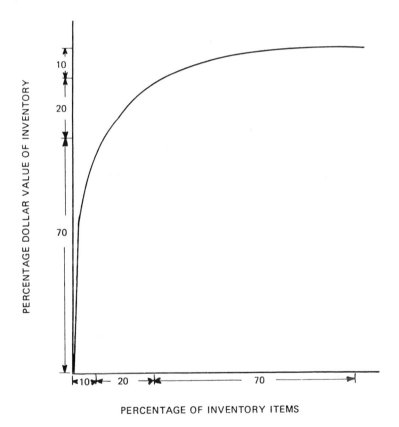

Figure 6.2 Pareto's law.

Class B. These items represent perhaps 20 to 30% of the items but about 25% of the total investment.

Class C. These items represent maybe 60 to 70% of the items and about 10% of the investment.

On Class A items, which have a high inventory value, ordering should be based on economic ordering quantity and close control of the inventory should be maintained. Minimum buffer stocks will be maintained in an effort to keep investment low.

Larger buffer stocks can prevail on the less expensive Class B items. Larger ordering quantities will prevail on Class B items than Class A since the cost of possession of these items will be less.

Since Class C items will represent only about 10% of the yearly inventory investment, very little control will be exercised on these items. The procedure here will be to maintain a buffer stock to accommodate a reasonable period such as 10 weeks. Then periodically, perhaps every six months, reordering can take place.

RECORD PROCEDURE

Whether computer-controlled or manual procedures are employed, there must be informative inventory records to assure that parts and materials are available for routine maintenance, repairs, and overhauls. Under a manual system, an inventory file may involve a visible card system, where the top portion of the card provides information relative to the item including its number, drawing number, description, vendors, and information about its usage. The lower part of the card can be an in-out balance record and can be set up so that it visually identifies when it is time to order material, when stock or stores are getting dangerously low, and when inventory is becoming excessive (see Fig. 6.3).

Different vendors supplying the same material usually will have their own identifying part number. These vendors and their respective part numbers should be identified on the inventory file, and parts from all vendors should be stored together thus avoiding duplication of stocked parts.

When purchasing renewal parts, it is usually wise to buy from the original supplier of the equipment. The original supplier will be able to provide components of the same quality as the original equipment and will usually give good service in the maintenance of the equipment they produce.

PT. NO.	DESCRIPTION					DWG. NO.

SPECIFICATIONS. _____

ECON. ORDER QTY _____ ANNUAL USAGE _____ MAX. INV. _____ MIN. INV. _____

VENDOR INFORMATION

VENDOR	PT. NO.	DWG. NO.	PRICE	DATE	REMARKS

RECEIVED			C O D E	WITHDRAWALS			BAL. ON HAND	RECEIVED			WITHDRAWALS			BAL. ON HAND	LOCAT.
P.O.	DATE	QTY.						P.O.	DATE	QTY.	REQ. NO.	DATE	QTY.		

Figure 6.3   Inventory file.

CENTRALIZED OR DECENTRALIZED STOREROOMS

The inventory of maintenance materials, in order to be of maximum worth to the goals of maintenance, should be readily available, in good condition, and be correctly identified. These criteria invariably bring up the question of whether centralized or decentralized storerooms will best fulfill the objectives of a sound engineering maintenance management program.

Centralized storerooms have the obvious advantages of reduced control records and labor costs for maintaining control and issuing stores and stock. On the other hand, decentralized storerooms allow the inventorying of parts in close proximity to where they will be needed, resulting in time savings in performing maintenance work.

Certainly in the vast majority of small and medium-sized plants the storage of maintenance inventory material should be done in a main storeroom that is centrally located. Under the control of the centralized storeroom may be field warehouses where dangerous or extremely heavy or bulky materials are kept. For example, very large forgings and castings that are spare parts for heavy machinery would not be transported to a centralized storeroom for inventorying. They would be stored near their use point.

In large plants occupying several buildings that may be some distance apart, it is desirable to have decentralized storerooms for high-use materials such as light bulbs, fuses, lubricants, screws, nuts, bolts, cotters, drive belts, and so on. For seldom used large heavy parts, the delivery of materials to job sites by personnel other than the performing craftsmen will often be desirable. By using the telephone or radios, this service can be directed so that maintenance crews can be instructed to proceed from one job to the next without returning to the central shops. Delivery trucks operated by material expeditors under control of the storeroom manager may be used for pickup and delivery of spare parts, materials, and supplies to the job sites. Of course, this effort must be scheduled effectively so that materials are available at the time the maintenance crew arrives at the work site.

STORAGE METHODS

Stores (finished components) should be stored in a separate area from stock (supply material for crafts). Stock supplies should be stored near the craftsmen's shops where they will be used. Thus lumber would be stored near the carpentry shop, pipe and tubing near the plumber's shop, electrical conduit near the electrician's shop, and so forth.

Storerooms and stockrooms should be designed in accordance with good plant layout practice. Heavy, bulky items should be stored as near their point of usage as possible and at a level where they can be easily reached and transported. Where mechanical equipment is involved in their transportation, there must be adequate aisle space for getting equipment to and from storage. For hand-operated fork trucks, this is usually 6 ft, but can be up to 8 ft depending on the load being transported, and up to 14 ft for a 6000-lb fork truck.

To conserve space, storage should be carried out to 24 in. below sprinklers. This will involve the use of shelves and bins, stacking, and overhead carriers.

Aisle space within the storeroom should be ample so that the attendant can easily acquire and dispense items. Thirty inches will allow room for two persons to pass.

All stocks and stores should be classified by both usage and bulk and stored accordingly. Those items with high usage should be readily accessible. They should be close to their point of dispension and use. Similarly, heavy or difficult to handle parts should be stored in an area and at a level for quick and convenient release. Another point to be considered in the specific location of stores and stock is to provide safeguards to protect the inventory from theft, deterioration, and spoilage. This means some items should be protected from moisture occurring in areas such as sweating pipes, damp floors, and leaking roofs. Others will need to be isolated from dust and dirt or from heat or cold. Special care will need to be observed to protect some items from pilferage through the use of locked cabinets or closed storerooms and stockrooms.

TWO-BIN INVENTORY CONTROL

On the less costly and frequently used Class B items and the vast majority of Class C items, control can be maintained using two-bin inventory control. After an economic order quantity has been determined, parts have been purchased, and then received; they are inventoried in two separate bins wherein each bin contains about the same number of components. On small items such as screw machine parts, this division can be readily determined by "weigh" counting.

As inventory is requisitioned to fulfill orders, parts are removed from one of the two bins. When this first bin becomes depleted a new order is placed. The inventory stored in the second bin is adequate to meet demands during the interval between placing of an order (either on a supplier or on the production facilities) and the time at which the order parts are delivered to the storeroom to replenish the two bins.

SAFETY STOCK AND LEAD TIMES

As indicated in the preceding section, lead time is the interval between the placing of an order and the time when the stock or stores are available for requisitioning to fulfill a maintenance work order. Although the demand for supplies in maintenance work is seldom constant, a planning model assuming linearity will usually permit the computation of safety stocks that will be representative of a sound cost effective policy. For example, by referring to Fig. 6.4 we note the reorder level BO, after the exhaustion of bin "A" (BC) is of sufficient magnitude that the demand based upon average requirements will not exhaust the inventory before the replacement order is fulfilled. Based upon the mean demand, AO provides an excess of inventory during the lead time DE. This excess of inventory is referred to as "safety stock."

When the inventory falls to level B at approximately time D, an order to replace inventory in bin A will be placed. The quantity of inventory required

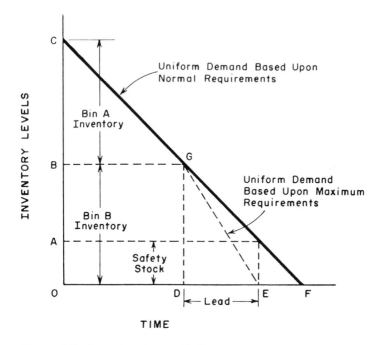

Figure 6.4   Operation of a two-bin inventory system.

will take care of the maximum demand during the lead time DE plus the safety stock OA, sometimes referred to as buffer stock. Using the demand based upon maximum requirements to determine the allowed lead time provides for alterations in the uniform demand premise that may make greater inventory demands during some particular time interval. Typically this maximum demand is based upon two standard deviations added to the mean demand. The safety stock AO should be of such magnitude to assure adequate service when a significant change takes place in either the lead or demand.

ECONOMIC ORDER QUANTITY

In the computation of the economic order quantity, we must take into consideration the cost of acquisitioning (ordering) and the cost of possessing (carrying) as mentioned previously. Ordering cost, as the name implies, is the cost of ordering. This cost is independent of the size of the order. When ordering parts internally that must be made in the plant, the ordering cost usually will include the various setup costs. It should be apparent that the unit cost of ordering decreases markedly with an increase in the order size as illustrated in Fig. 6.5. If the cost of ordering C_o and the order quantity is a and the periodic usage is U, then the cost of ordering per usage time (1 yr, or other increment of time) is

$$C_u = \frac{C_o U}{a}$$

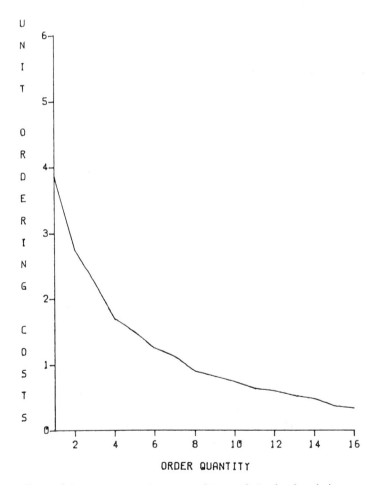

Figure 6.5 Relationship between order quantity and unit ordering costs.

The cost of possession (or carrying cost) is made up of two costs. First is the cost of the monetary value of the inventory that is tied up. This includes current interest rates plus any allowance for inflation or decrease in value of the dollar. Second is the cost of physical storage, which is usually based upon cubic foot occupance. The rate assigned per unit of occupance takes into consideration building cost, depreciation, heat, lighting, wages of stock clerk, insurance, and so on. Typically the costs of possession are handled as a percentage of the purchase cost. Figure 6.6 illustrates the linear relation between order quantity a and the possession costs per unit of time, such as a year. For example, if C_p is the unit purchase cost and the cost of possession is a percentage of C_p for some unit of time which can be identified as i, then C_{pi} is the cost of possession for the unit of time on which i is based. Assuming a linear relation in the withdrawal of inventory and a order quantity of a, the average inventory would be $a/2$ and the possession cost per unit of time would be:

$$C_{pi}\frac{a}{2}$$

Then the total costs per unit of time, say 1 yr, would equal

$$C_o\frac{U}{a} + C_{pi}\frac{a}{2}$$

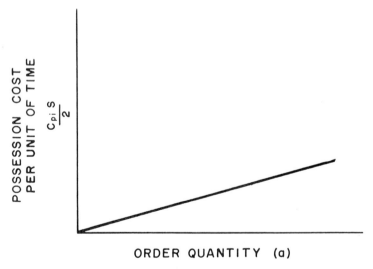

Figure 6.6 Relationship betwen order quantity and cost of possession.

And the minimum cost associated with order quantity would be

$$a_{\min} = \sqrt{\frac{2C_oU}{C_{pi}}}$$

This order quantity that provides a minimum cost is often referred to as Q (see Fig. 6.7).

PHYSICAL INVENTORY

The reorder point equation that is typically used for maintenance stores and stock is

$$G_A = B + UL$$

where

G_A = reorder point in units (Class A items)

B = buffer stock in units

U = average inventory per week

L = average lead time in weeks (see Fig. 6.4)

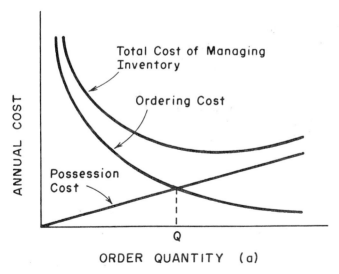

Figure 6.7 Relationship between order quantity and total cost of managing inventory.

This equation is based on the company maintaining a perpetual inventory record system. This is recommended on all Class A items and may be used on those Class B items that have high usage. In these items, a new order is placed as soon as an item's inventory level drops to the reorder point.

Where perpetual inventory records are not maintained, such as in some Class B and Class C items, the preceding equation will need to be modified to allow for the decline in inventory between the time when reordering would have taken place if there were a perpetual inventory control and the time the next physical count is made. Here the preceding equation becomes

$$G_{B\&C} = B + U\left(L + \frac{T}{2}\right)$$

where

$G_{B\&C}$ = reorder point in units (Class B and C items)

T = time between physical counts in weeks

A physical inventory count of all maintenance stock and stores should be made periodically. This actual count will provide a check on the control systems being utilized for all items, whether of Class A, B, or C. It will reveal if any pilferage is taking place, if the quality of the stock and stores is being maintained, if the inventory is being adequately protected from the elements, and will identify low- and high-demand items.

Where inventory is controlled by perpetual inventory records, a physical count once a year is considered satisfactory. A perpetual inventory system, sometimes referred to as a continuous review system, implies that the status of the inventory is checked each time a disbursement is made. If the reorder point has been past, a new order is issued immediately.

Those Class B and C items for which the reorder point is determined by the periodic physical count usually have a 3-mo. period. Those Class C items controlled by the two-bin system would be physically inventoried annually.

BAR CODING

Bar coding is an important method for managing inventories. The black bars and white spaces represent ten digits that identify both the item and the manufacturer. Once this Universal Product Code (UPC) is scanned by a reader, the decoded data can be sent to a computer that records timely information such as inventory status.

The following reasons illustrate the desirability of incorporating bar coding with maintenance inventory control:

1. *Accuracy.* Less than 1 error in 3.4 million characters is a representative performance (U.S. Army LOGMARS study). This compares favorably with the 2 to 5% error that is characteristic of keyboard data entry.
2. *Performance.* A bar code scanner enters data three to four times faster than typical keyboard entry.
3. *Acceptance.* Most employees enjoy using the scanning wand. Inevitably, they prefer using a wand to keyboard entry.
4. *Low cost.* The cost of adding this identification to inventory items is extremely low.
5. *Portability.* An operator can carry a bar code scanner into any area of a plant to determine inventories, status of a maintenance order, and other information.

The typical maintenance inventory storage bin label provides the following information: part description, department number, storage number, basic stock or stores level, and order point. Considerable time is saved by using a scanning wand to gather all this detail from a bin for inventory recording.

COMPUTERIZED MAINTENANCE INVENTORY CONTROL

With the memory and speed of computation of the computer, it is often advantageous to introduce an on-line interactive computerized system tailored to meet the specific needs of maintenance parts inventory management. Usually a computerized system is advantageous when inventories exceed 5000 parts and the number of inventory transactions is substantial. Before deciding on the software/hardware package that will best fit a company's or business' specific requirements, a survey should be made that will provide information as to the number of pieces of equipment to be serviced, the number of spare parts and different stocks maintained in inventory, and the average number of transactions that typically take place every month.

The software utilized or developed will generally include three modules with each module having its own objectives:

1. *Stockroom and storeroom.* The stockroom and storeroom personnel should at all times be able to service the supply needs of the tradespeople by rapidly assessing the spare parts and materials inventories and furnishing the needed supplies in an efficient manner.
2. *Craft- and/or tradesperson.* The maintenance mechanic should be able to expeditiously repair, overhaul, and maintain all plant facilities by having repair parts and materials easy to both requisition and

identify. A feature of this module is the ability to assess inventory stock and stores by part number, description, equipment number, manufacturer's number, and work order number.

3. *Inventory control.* Inventory levels should be maintained that keep inventory levels financially reasonable while avoiding stockouts. Reorder point formulas will be modified to accommodate changes in actual usage and the actual lead times. Computer-generated reports will keep management abreast of the total system. In Chapter 10, computerized systems are discussed in more detail.

SUMMARY

To maintain high plant and facility availability, there is a tendency to overstock spare parts and maintenance materials. Excess inventory is a costly luxury that business and industry cannot afford. Inventory sizes should be based on careful analysis. The alternative of repair as opposed to replacement should always be considered, not only to reduce spare parts inventory, but also to provide greater plant and facility availability.

Once usage lead times, availability, costs, interest rates, storage costs, inflation, and chance for spoilage have been taken into consideration, then economic order quantities should be determined and inventory control procedures should be incorporated. Those items that represent the greatest monetary value, Class A, should be controlled closely and those items that involve the least capital investment, Class C, require the least control.

Stores and stocks should be inventoried in separate storerooms. Good plant layout should be incorporated in all storerooms and stockrooms. There should be ample aisle space to accommodate the material-handling equipment that will be utilized.

Computerized inventory control is cost-effective when the number of inventory transactions as well as the number of items carried in inventory become reasonably large.

SELECTED BIBLIOGRAPHY

Lewis, C. D. *Demand Analysis and Inventory Control.* Farnborough, England: Saxon House, 1975.

Peterson, P. A., and Silver, E. A. *Decision Systems for Inventory Management and Production Planning.* New York: Wiley, 1979.

Tersine, R. J. *Production/Operations Management: Concepts, Structure, and Analysis.* New York: North Holland, 1980.

Young, T., and Fu, K. S. *Handbook of Pattern Recognition and Image Processing.* San Diego: Academic, 1986.

7
Maintenance Planning and Scheduling

Planning and scheduling are vital ingredients of a successful engineering maintenance program. The vast majority of maintenance work that is performed should be planned and scheduled. Only emergency repairs are made without advance planning scheduling. All maintenance work should be planned so that the quality and cost-effectiveness of the work is assured. Emergency work must be planned as it is taking place, operation by operation.

From the standpoint of planning and scheduling, maintenance work may be classified as follows:

1. *Emergency repairs.* Maintenance work may be requested in this category verbally by any company employee. This is usually done by telephone. This request should be made to the maintenance supervisor or plant engineering manager. A confirming work order request will be initiated as soon as convenient after the verbal request. Maintenance schedules are interrupted to take care of the emergency repairs, since this work is not scheduled in advance.
2. *General repairs, overhaul, and replacement.* Work order requests are made by section foremen. All general repairs should be planned and scheduled.
3. *Preventive maintenance.* Preventive maintenance is planned and scheduled. Those facilities where preventive repair will result in the

125

highest return on investment will receive high priority in the scheduling accordingly.

4. *Routine*. Routine maintenance are those maintenance operations of a periodic nature such as routine lubrication, inspection, and minor repetitive jobs estimated to require a small amount of time to accomplish (i.e., less than 2 hr). This work often is covered by a blanket order. These work orders are usually initiated by the maintenance department.

PLANNING

In Chapter 2, it was brought out that planning is one of the essential programs of a modern maintenance management system. Planning is integral with scheduling.

Upon receipt of the request for a work order (either written or verbal), the maintenance planner will develop all those details included on the work order that will result in the job being done with high quality at a minimum cost and at a time compatible with the urgency of the request and other scheduled maintenance work. Thus planning includes all those functions related to the origin of the work order, including the preparation of necessary drawings, sketches, bills of material, purchase requisitions, labor planning sheets including standard times, and any other data required preliminary to the release of the work order for the actual accomplishment of the work. Of course, in the case of emergency work such as power failure there will be insufficient time for planning to take place in advance of the work order preparation. Repair work must be begun at once with the planning taking place in concert with the various steps involved in the repair operation.

The maintenance work order will not provide enough space to carry the planning detail characterized by extensive repair, overhaul, and plant expansion projects. In these cases (where the size of the project involves more than 20 hr of maintenance time), it is good practice to complete a maintenance planning sheet (see Fig. 7.1).

In order to do an effective job of planning, it is generally wise for the maintenance planner to visit the location where the work is to be performed to assure reliability of information as to accessibility of the facility at the work site, the type and condition of the floor, the location of the service lines (power, air, water, drains, etc.), the lighting, the location of columns or other supports. It is also a good idea for the planner to inspect the involved installation equipment. With all these details clearly in mind, the planner is in a much better position to specify the best sequence of operations for performing the

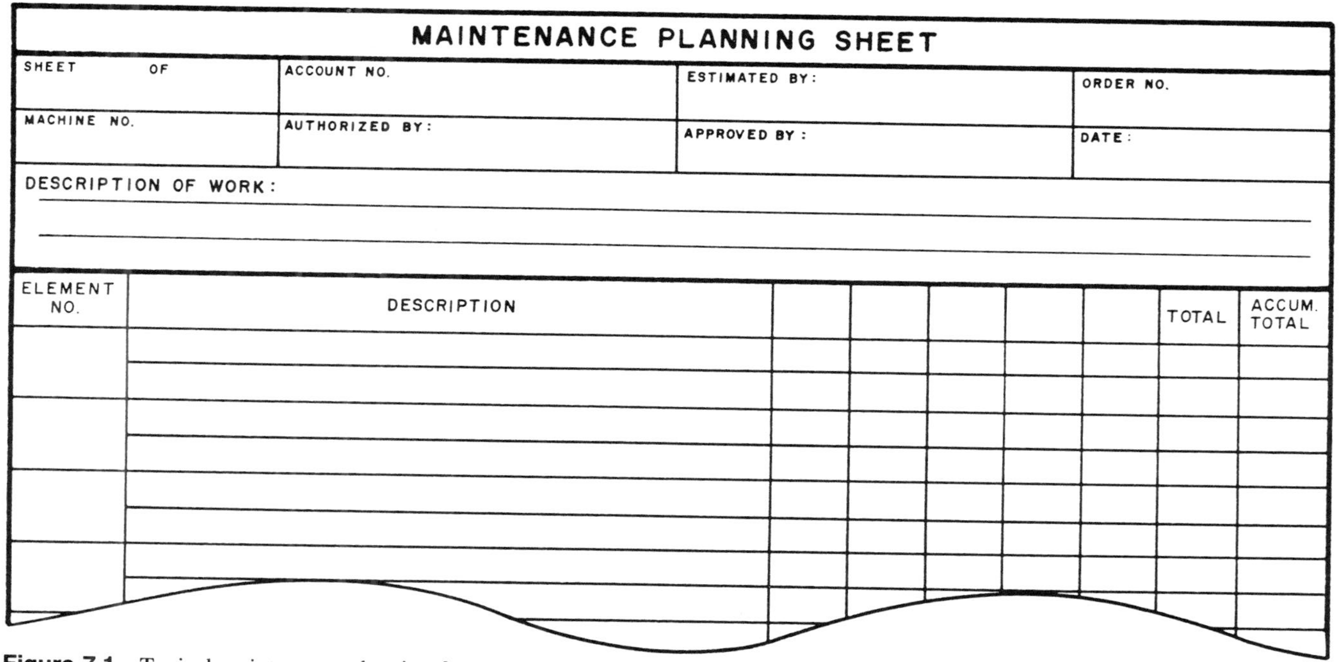

Figure 7.1 Typical maintenance planning form.

work and identifying the type and amount of craft labor and materials needed to do the job.

In completing the maintenance planning form, each work element should be clearly defined so that the maintenance workers will know exactly what is required. Any special tools, materials, or equipment that are needed will be listed in the space provided.

Two spaces have been provided under each craft or work element. In the upper space, the planner indicates the number of workers by craft that should be assigned to the job, and the lower space is for the estimated total hours by craft.

The total hours for each element and for the job are summarized in the spaces provided. On occasion, it may be desirable to follow a different procedure than what has been outlined on the maintenance planning form. This can be due to a variety of reasons. Upon beginning the repair or overhaul, it may become evident that either more or less work than anticipated needs to be done. Or the size of the job may increase as it develops. Additional repairs might best be done while the facility is down. There may be a change in design or a change in equipment location; there could be an improved work procedure that was not evident when the planning took place; oversights may have been made by the foremen or individual requesting the repair; outright errors may have been made. When it becomes necessary to change the work procedure as outlined on the maintenance planning sheet, a maintenance planning change notice form should be initiated and processed (see Fig. 7.2).

In completing the planning and the planning change forms, the maintenance planner should feel free to draw on the background of all expertise that is available. The pooled knowledge of the maintenance planner, maintenance foreman, craftsmen, and plant engineer will inevitably be superior to the judgment of any one of these parties.

SCHEDULING

Scheduling maintenance work can be handled in three degrees of refinement: (1) long-range or master scheduling, sometimes referred to as work ahead schedule; (2) weekly scheduling—the maintenance and repair work scheduled for the current week; and (3) daily work scheduling—the maintenance and repair work to be completed each day.

General repair and overhaul, preventive maintenance, and routine maintenance are all scheduled. Jobs which must be done as emergencies arise are planned for by providing a percentage of maintenance plant capacity time to all three schedules: long-range, weekly, and daily. For example, from past

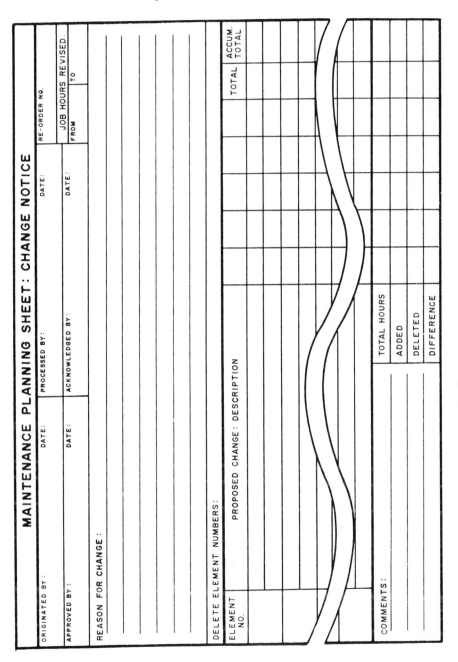

Figure 7.2 Typical planning sheet–change notice form.

experience 10% of maintenance activity may have taken place in conjunction with emergency repair work. If thirty maintenance employees comprise the labor force, then the master schedule will reflect 120 hr every week as being reserved for emergency work (30 × 40 hr/wk × 0.10 = 120 hr). It should be understood that the craft distribution of this 120 hr would reflect the expectations of what will be needed.

Long-Range Scheduling

Long-range scheduling is based upon the existing maintenance work orders including the blanket orders issued for routine and preventive maintenance and the anticipated work brought about through emergency repair. Long-range scheduling is frequently carried out on a craft basis so that a balance of manpower will be maintained. It will reveal when it will be necessary to add to the maintenance work or subcontract a portion of the maintenance work. Similarly the master schedule will bring out when it is desirable to retrench the maintenance work force as a result of a declining volume of work. When overhaul maintenance volume diminishes, there must be evidence to determine the actual number and skills of craftsmen needed to perform the reduced load. Without firm scheduling there will be a tendency for the entire maintenance work force to slow down so that the available work orders will last. Unless we continually balance our maintenance work force with the available volume of work, the cost of maintenance work will rise progressively.

In a growing operation, the amount of maintenance work necessary will also grow. The long-range schedule will bring out the need to add manpower to the force and will clearly indicate when it will be needed. Figure 7.3 illustrates how overall maintenance capacity may be scheduled in an expanding or growth situation.

It should be understood that long-range schedules will need to be revised regularly to accommodate changes in plans. More or fewer emergency repairs than anticipated can result in significant changes in the long-range schedule.

The accuracy of the long-range schedule is based on the thoroughness of the planning and the validity of the time standards that are applied. Without sound planning of general repairs, overhaul, and preventive maintenance work and control of routine maintenance, supplemented with standards based on measurement, the reliability of the master schedule will be poor. Table 7.1 illustrates a typical long-range schedule by craft.

Weekly Scheduling

Weekly or periodic scheduling is the function that provides for the economic accomplishment of work at a time that maximizes the objectives of the com-

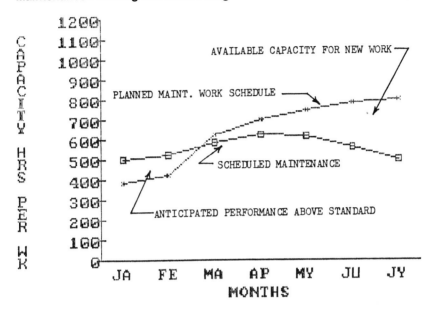

Figure 7.3 Planned and scheduled maintenance work.

pany. The weekly schedule is constructed from the long-range schedule and provides for the orderly introduction of emergency work as it occurs. Typically, approximately 85% of the weekly schedule is based on general repair, overhaul, preventive, and routine maintenance work. Fifteen percent of the workweek is held open for emergency work. It can be understood that this 0.85-to-0.15 ratio will not prevail week after week. There will be times when perhaps as much as one-third of the week or more will be utilized in performing emergency work. When this happens the next week's schedule will have to be modified in order to complete unfinished work from the current week's schedule. On occasion, it may be necessary to schedule a portion of the maintenance crew on an overtime basis in order to handle the emergency work so that needed production facilities can be returned to operable condition.

If emergency work is occurring so frequently that it is regularly utilizing more than 15% of the available maintenance work force, an effort should be made to determine the causes. An ongoing program of modern maintenance engineering includes improving the reliability of all production equipment. It also includes, through predictive and preventive maintenance, the keeping of all equipment in good operating condition and the planning for equipment overhaul well in advance of breakdown.

Table 7.1 Long-Range Maintenance Schedule

Craftsmen	March Order no.	hr	April Order no.	hr	May Order no.	hr	June Order no.	hr	July Order no.	hr
Machinists	Emerg. Est.	200	Emerg. Est	200	Emerg. Est.	200	Emerg. Est.	200	Emerg. Est.	200
10	Blanket	300	Blanket	300	Blanket	300	Blanket	300	Blanket	300
1730 hr/mo	475216	145	475185	150	479528	22	479539	250	479543	80
	478401	425	477192	55	479531	125	47479540	150	479544	20
	478514	660	478514	125	479532	250	479542	180	479546	12
			478518	600	479538	450				
			479520	300						
Millwrights	Emerg. Est.	160	Emerg. Est.	160	Emerg. Est.	160	Emerg. Est.	160	Emerg. Est.	160
8	Blanket	240	Blanket	240	Blanket	240	Blanket	240	Blanket	240
1384 hr/mo	475195	175	478401	90	479532	200	479539	200	479580	400
	475198	150	478422	200	479538	400	479540	125		
	475204	40	478417	412			479542	650		
	478514	619	478429	16						
			478435	266						

Craftsmen	March Order no.	hr	April Order no.	hr	May Order no.	hr	June Order no.	hr	July Order no.	hr
Electricians	Emerg. Est.	100	Emerg. Est.	100	Emerg. Est.	100	Emerg. Est.	100	Emerg. Est.	100
4	Blanket	140	Blanket	140	Blanket	140	Blanket	140	Blanket	140
692 hr/mo	475216	40	478422	120	479532	100	479539	120	479539	120
	478401	122	478417	212	479538	180	479540	8	479612	40
	478514	240	478435	80	479544	20	479052	320	479601	80
	478519	50	478440	40						
Plumbers	Emerg. Est.	100	Emerg. Est.	100	Emerg. Est.	100	Emerg. Est.	100	Emerg. Est.	100
5	Blanket	160	Blanket	160	Blanket	160	Blanket	160	Blanket	160
863 hr/mo	475216	80	478422	113	479532	40	479539	80	479580	140
	478401	200	478417	220	479538	200	479540	48	479601	100
	478514	320	478418	20	479542	16	479548	24	479612	20
	478542	3	478430	16	479544	30	479550	120	479614	80
			478432	234						

In the weekly schedule, the planning function will not only provide for the work to be completed in the current week, but also for the following week. Thus there will always be a backlog of two weeks of work orders scheduled according to priority. If, during a period of several days, no emergency work is requested and consequently the maintenance force is well ahead of schedule, there will always be planned work ahead so all maintenance employees can continue working at a normal or better than normal pace.

When planning the sequence of the work order scheduled for the week, consideration must be given not only to the nature of the repair in connection with the production equipment, but also to the availability of special components and stock to perform the repair. Also consideration should be given as to the availability of specialty craftsmen with unique skills that may be required to do the job. After all these factors are evaluated, a priority is assigned that places the work order in its most favorable sequence. Work that is scheduled in the weekly schedule should not be introduced into the current week until it has been verified that all material requirements are on hand and all personnel requirements are available. Those work orders that are scheduled in the current week are arranged in a priority sequence where the general repair and overhaul are customarily performed first, preventive maintenance second, and routine maintenance last. Of course, when emergency repairs occur they are introduced in the schedule in advance of all other work except other emergency repairs of higher priority that may be taking place. Of those general repair and overhaul work orders that are scheduled during the week, an order of priority is based upon requested completion date, date when order is received, and most important the impact that the facility being repaired has on the production output of the company or business.

It is important that the weekly schedule be realistic. Assuming the maintenance department achieves standard performance, the scheduling should be based upon normal (100%) effort. Similarly the scheduled hours should provide an allowance for typical absenteeism which may run as high as 10%. Table 7.2 illustrates a weekly schedule by craft.

Daily Work Scheduling

The daily work schedule is constructed from the weekly schedule and is usually provided by maintenance supervision the day before the scheduled day. Frequently, for a variety of reasons, changes in the planned weekly schedule will need to be made. Emergency work may be introduced or removed, some planned work may take more or less time than anticipated, shortages in materials or parts may be noticed, a higher rate of absenteeism

Table 7.2 Weekly Maintenance Schedule

Week of June 8

Craftsmen	Monday Order no.	hr	Tuesday Order no.	hr	Wednesday Order no.	hr	Thursday Order no.	hr	Friday Order no.	hr	Satuday Order no.	hr	Sunday Order no.	hr
Millwrights	466781	16	466781	16	466781	8	355812	8	563271	8				
	355812	16	355812	16	355812	16	563271	16	917248	4				
	355818	4	284712	2	471408	4	917248	8	571410	8				
	700410	8	296041	4	563271	12	697116	8	571411	8				
	814112	8	814112	10	917248	8	867414	4	867414	4				
	274406	4	45416	4	724018	6			924810	8				
			62408	4	827011	2								
Pipefitters	466781	8	466781	8	466781	8	355812	4	563271	4				
	355812	8	355812	8	355812	8	700716	4	571410	4				
	356414	4	814112	4	700410	4	852910	8	571411	4				
	463281	2	700410	8	700716	8	868112	8	924810	2				
	814112	6	700716	4					967341	8				
	700410	4												

135

Table 7.2 Continued

Week of June 8

Craftsmen	Monday		Tuesday		Wednesday		Thursday		Friday		Saturday		Sunday	
	Order no.	hr	Order no.	hr	Order no.	hr	Order no.	hr	Order no.	hr	Order no.	hr	Order no.	hr
Electricians	466781	12	466781	12	466781	4	652418	8	652418	8	652418	8	652418	4
	355812	4	355812	4	355812	4	852910	4	852910	4				
	814112	4	814112	4	814112	8	868112	4	563271	2				
	797412	2	937112	8	937112	8	714886	8	967341	6				
	884666	2	754482	4	767451	4	452771	2	854624	4				
	937112	8			974812	4	561419	2	568172	2				
							854624	4	666412	6				
Machinists	466781	24	466781	16	466781	8	652418	12	652418	8				
	355812	16	355812	16	355812	8	852910	8	852910	8				
	700410	8	700410	4	937112	16	868112	4	967341	8				
	814112	8	814112	2	974812	8	714886	8	854624	4				
			937112	4	767451	4	452771	4	756418	12				
			754482	8	854624	4	854624	8	666412	8				
							561419	2	972114	4				

than planned may have occurred, and so on. The daily schedule includes only those work orders for which materials are on hand and the workmen are available to handle the assignment. As mentioned earlier, a priority is assigned to the daily schedule. Three levels are considered: scheduled, rush, and emergency.

From this daily schedule, the area maintenance supervisor is able to assign each of his individual craftsmen work for the following day. In this way each maintenance employee knows in advance his work assignment and the time estimated for completion. All other pertinent details such as materials, special tools, and equipment required is obtained from the work order.

Short Interval Scheduling

An alternative to the daily work schedule is the short interval scheduling technique. Here the maintenance supervisor assumes the responsibility for the daily scheduling of all those craftsmen and workers coming under a particular jurisdiction. However, the workers' schedule is based on a short interval, which is frequently on a 4-hr basis rather than on a whole 8-hr workday. The supervisor will use the weekly or long range schedule as the basis for establishing the short interval schedules.

It can be appreciated that under short interval scheduling, the supervisor will be more concerned with the progress of all workers coming under his jurisdiction than if the scheduling were done centrally. Since he scheduled the work, he will be more familiar with what needs to be done and the methods that should be utilized for the successful completion of the work. With the immediate supervisor responsible for the work scheduled, there is more incentive for the maintenance worker to meet the schedule even if the scheduling is tight (i.e., overscheduling in the allotted time). To be successful, short interval scheduling requires more follow-up and time at the work site on the part of the supervisor. This additional follow-up tends to assure the quality of the maintenance work done and increase the productivity of the maintenance craftsmen.

The principal reason for the short interval of scheduling is to maintain control and permit adjustments in the schedule as work progresses. In some cases the supervisor will schedule his workers in 2-hr intervals. For maintenance work however, the 4-hr interval is generally more useful.

Requirements for Sound Scheduling

No matter what the level of scheduling is—long-range or short interval—there are several requirements that must be maintained if the scheduling is to be successful.

1. There must be a time standard system based upon measurement. Thus standards need to be developed in advance of the time that the work is done. These standards may be developed either through slotting techniques or standard data.
2. There needs to be a written work order system that explains accurately the work to be done, the methods to be followed, the crafts to be employed, and the desired completion date.
3. There must be information available as to the capacity for work by craft. Thus there should be current information as to the number of employees for each shift identified by craft or crafts that they are able to perform. This manpower availability report should reflect the anticipated man-hours lost by absenteeism, vacations, and so on.
4. There must be information provided as to the availability of stock and stores. Also there needs to be information on promised dates for needed spare parts, supplies, tools, materials, and so forth.

Other information that the scheduler will use in developing reliable schedules is a summary of those jobs that are behind schedule as well as the plant production schedule. The overdue maintenance work will get high priority in the development of the current schedule. Having access to the plant production schedule will let the maintenance scheduler know when scheduled facilities must be available for production. It will also let the scheduler know when the facilities may be serviced without interrupting the production schedule.

Program Evaluation Review Technique

A useful scheduling tool for major equipment overhauls involving several crafts, plant expansions and modifications, and similar large projects where many activities are performed in concert is the program evaluation review technique (PERT). PERT charting has been defined as "a prognostic planning and control method that graphically portrays the optimum way to attain some predetermined objective generally in terms of time."* Usually the maintenance planner is able to use PERT charting to improve scheduling from both the standpoint of cost reduction and completion time of the project.

In using PERT for scheduling, the maintenance planner will generally provide two or three time estimates for each activity. If three time estimates, they are based upon the following questions:

1. What is the time in which you can expect to complete this activity if everything works out ideally (optimistic estimate)?

*B. W. Niebel. *Motion and Time Study*, 9th ed. Homewood, IL: Richard D. Irwin, 1993.

2. Under normal conditions, what would be the most likely duration of this activity? (This is the estimate based on slotting or standard data.)
3. What is the time required to complete this activity if almost everything goes wrong (pessimistic estimate)?

With these estimates, a probability distribution of the time required to perform the activity can be made. On the PERT chart, events represented by nodes are positions in time showing the start and the completion of a particular operation or a group of operations. Each operation or group of operations in a maintenance department is referred to as an activity and is identified as an arc on the PERT chart. Each arc has attached to it a number representing the expected time (days, weeks, or months) needed to complete the activity. Activities that utilize no time or cost, though necessary to maintain a correct sequence, are referred to as dummy activities and are plotted as dotted lines.

The minimum expected time needed to complete the entire project would correspond to the longest path from the initial node to the final node. For example, in Fig. 7.4, the minimum expected time needed to complete the project would be the longest path from node 1 to node 12. The longest path is termed the critical path since it is this path that establishes the minimum expected time to complete the project. There is always at least one such path through any project. However, more than one path can reflect the minimum expected time needed. This is the meaning behind the concept of critical paths.

It should be evident that activities which do not lie on the critical path have a certain time flexibility. This time flexibility, or freedom, is referred to as *float*. The amount of float is computed by subtracting the normal time from the time available. Thus, the float is the amount of time that a noncritical activity can be lengthened without delaying the project's completion date.

Figure 7.4 illustrates an elementary network portraying the critical path. This path, identified by a heavy line, would involve a duration of 27 weeks. Several methods can be used to shorten the project's duration. The cost of the various alternatives can be estimated. For example, let us assume that the cost table shown in Table 7.3 has been developed and that a linear relation exists between the time and the cost per week.

The cost of various time alternatives can be readily computed. For example, Table 7.4 illustrates how this project could be reduced in time from 27 weeks to 18 weeks at an added cost of $7800.

Job Assignment Schedule

The job assignment schedule is a daily schedule that identifies the maintenance employees who will be working on specific maintenance orders. It is typically

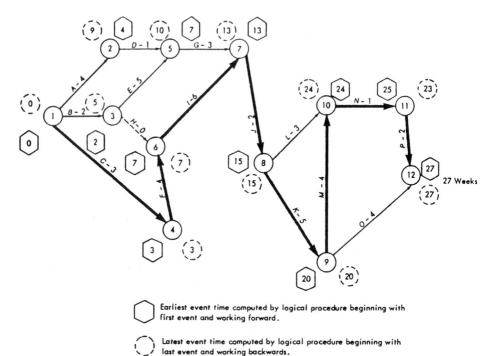

Earliest event time computed by logical procedure beginning with first event and working forward.

Latest event time computed by logical procedure beginning with last event and working backwards.

Figure 7.4 An example of a PERT chart. Code numbers within the nodes signify events. Connecting lines with directional arrows indicate operations that are dependent upon prerequisite operations. Time values on the connecting lines represent normal duration in weeks. Hexagonals associated with events show the earliest event time. Dotted circles associated with events show the latest event time. The heavy line indicates the critical path. (Courtesy of B. W. Niebel, *Motion and Time Study*, 7th ed., Homewood, IL: Richard D. Irwin.)

released late in the day preceding the scheduled day's work. The job assignment schedule is often developed by the maintenance foremen in concert with the maintenance planner. Its objective is to assure that work scheduled for a given day will be begun, that unnecessary instruction time or waiting time for assignments will be eliminated, and that the foremen will maintain control of those reporting to him by having up-to-date knowledge of where his craftsmen are working and what they are doing. This schedule is usually placed in a conspicuous place in the maintenance office (see Table 7.5).

Table 7.3 Normal and Emergency Costs of Various Activities

Activities	Normal Weeks	Normal Dollars	Emergency Weeks	Emergency Dollars
A	4	4000	2	6000
B	2	1200	1	2500
C	3	3600	2	4800
D	1	1000	0.5	1800
E	5	6000	3	8000
F	4	3200	3	5000
G	3	3000	2	5000
H	0	0	0	0
I	6	7200	4	8400
J	2	1600	1	2000
K	5	3000	3	4000
L	3	3000	2	4000
M	4	1600	3	2000
N	1	700	1	700
O	4	4400	2	6000
P	2	1600	1	2400

Maintenance Employee Dispatch Board

The employee dispatch board is a form of the job assignment schedule. Its objective is to provide a procedure for the maintenance foremen to schedule work to specific employees well in advance of the time that the work will be done. Its purpose is to assure that there is always a work load ahead of each maintenance worker, thus reducing instruction and waiting time to a minimum while maintaining good performance on all work done.

Typically the dispatch board is maintained on one of the walls in the maintenance office. Each maintenance employee's name is shown, below which is a clip or a pocket for holding work orders. As work orders are received, the maintenance foremen will evaluate the work that is to be done and its urgency and then assign the job to qualified employees by attaching a copy of the work order to the employee's clip or pocket. The maintenance foremen will arrange the orders displayed on each employee's clip in order of priority. The most urgent work order will be displayed first; this will be followed by the second most important and so on.

Table 7.4   Comparison of Scheduling Alternatives Using Data from Table 7.3.

Alternative schedule	Cost ($)
27-week schedule—normal duration of project	22,500
26-week schedule—the least expensive way to gain one week would be to reduce activity M or J by one week for an additional cost of $400	22,900
25-week schedule—the least expensive way to gain two weeks would be to reduce activities M and J by one week each for an additional cost of $800	23,300
24-week schedule—the least expensive way to gain three weeks would be to reduce activities M, J, and K by one week each for an additional cost of $1,300	23,800
23-week schedule—the least expensive way to gain four weeks would be to reduce activities M and J by one week each and activity K by two weeks for an additional cost of $1,800	24,300
22-week schedule—the least expensive way to gain five weeks would be to reduce activities M and J by one week each, activity K by two weeks, and activity I by one week for an additional cost of $2,400	24,900
21-week schedule—the least expensive way to gain six weeks would be to reduce activities M and J by one week each and activities K and I by two weeks each for an additional cost of $3,000	25,500
20-week schedule—the least expensive way to gain seven weeks would be to reduce activities M, J, and P by one week each and activities K and I by two weeks each for an additional cost of $3,800	26,300
19-week schedule—the least expensive way to gain eight weeks would be to reduce activities M, J, P, and C by one week each and activities K and I by two weeks each for an additional cost of $5,000 (Note that a second critical path is now developed through nodes 1, 3, 5, and 7.)	27,500
18-week schedule—the least expensive way to gain nine weeks would be to reduce activities M, J, P, C, E, and F by one week each and activities K and I by two weeks each for an additional cost of $7,800 (Note that by shortening the time to 18 weeks, we develop a second critical path.)	30,300

The maintenance foremen should review the entire dispatch board regularly to assure that projected schedules are being realized. When the maintenance employees finish their current jobs, they will go to the dispatch board and obtain their next work order (that one appearing first in the clip under their name).

Table 7.5 Job Assignment Schedule

	Wednesday October 15				
Clock no.	Name	Assignment (order no.)	Clock no.	Name	Assignment (order no.)
150	S. Zook	652418	299	P. Zook	563271
222	G. Esh	852910	316	J. Laub	917248
					867414
256	L. Deeg	756418	382	R. Aumiller	924810
262	D. Buchanan	758418	394	B. Kauffman	571411
		972114			
274	E. Stringer	666412	438	G. Hoover	571410

The dispatch board technique will help assure that all work is performed in the desired priority and that each employee has an adequate backlog of work. An examination of the dispatch board provides information as to the need of corrective action whenever an excess or diminished workload develops.

Scheduling Using Computers

Computers can be helpful where maintenance scheduling of personnel is dynamic. It is always desirable to have a scheduling system that matches available maintenance work to personnel requirements and still gets the work completed when desired. Since both the work load and the available manpower by skill will vary markedly over time, the maintenance scheduling system will need to take the following steps:

1. Identify labor needs based on service levels, maintenance standards, personnel skill proficiencies, and vacation and sickness patterns.
2. After labor needs are identified by craft, develop them under the several alternatives that are most likely to exist in the future based on the existing production load, backlog of maintenance work, time of year, and so on.
3. Identify alternative means of matching personnel to maintenance skill job requirements. These include utilization of cross-trained employees in their skills, development of labor pools, and the use of part-time employees from either production or temporary recruits.
4. Regularly (daily) balance the availability of maintenance labor by skill with the projected work schedule.

The information system that maintains all the necessary data and rapidly provides the working schedule that optimizes movement of personnel of the maintenance department will invariably use the capabilities of computers. Since each information system is tailored to the unique requirements of the specific organization or business, at this time only the fundamental inputs for the system will be identified.

Weekly, the available employees (taking into consideration all anticipated absences) by craft will be identified. The work schedule, including anticipated emergency work orders, will be computed. This work schedule is based upon standard times and anticipated performance of all maintenance employees. If an excess or shortage of workmen exists, these will be identified as to craft and the number of hours that are short or in excess. If an excess of labor exists, work scheduled for the following week will be advanced. If a shortage of labor exists, then the individual in charge of scheduling will either utilize needed personnel from a labor pool or else modify the schedule so that it will be adequately staffed or arrange for overtime work in order to meet the desired schedule.

As work is completed, the system will be updated. Similarly, when any unplanned situation develops such as an influx of emergency work, the system will be updated.

SUMMARY

The vast majority of maintenance work, perhaps 85–90% should be planned and scheduled. This work load is made up of preventive and predictive maintenance, general repair and overhaul, replacement, expansion, and routine maintenance. The remainder of the maintenance work performed can be classified as emergency maintenance, which cannot be accurately scheduled in advance of the breakdown.

Planning of maintenance work should be carefully performed by experienced technicians. Planned work invariably is of better quality and is more reliable than work that is not planned. Furthermore, planned work usually will be performed in less time, it will allow better utilization of skilled craftspeople, and it will permit much more accurate scheduling.

Scheduling is usually performed at three levels: long range, weekly, and daily. All scheduling must provide a certain percentage of the available time to accommodate emergency repair of breakdowns. Daily schedules should be made only after all needed spare parts, supplies, and materials have been confirmed to be on hand. Only by scheduling maintenance work can the available work be balanced with the working force. Scheduled maintenance work not only provides for completing work when needed, but will result in work being performed in much less time than if the work is neither planned nor scheduled.

With modern data base management systems, utilizing the capability of the computer, all three levels of scheduling can be effectively carried out. Computerized scheduling is simplified since all necessary information needed to develop schedules may be rapidly retrieved from the data base.

SELECTED BIBLIOGRAPHY

Greene, J. H. *Production and Inventory Control: Systems and Decisions*. Homewood, IL: Richard D. Irwin, 1974.
Nakajima, Seiichi, Yamashina, Hatime, Kumagai, Chitoku, and Toyota, Toshio, "Maintenance Management and Control," In: Gavriel Salvendy, Ed. *Handbook of Industrial Engineering*, 2nd ed. New York: Wiley, 1992.

8

Preventive and Predictive Maintenance

PLANNED MAINTENANCE AND EQUIPMENT FAILURE

The terms preventive maintenance and predictive maintenance are often used interchangeably by those working in engineering maintenance management. In both preventive and predictive maintenance, we are able to plan and schedule the maintenance, therefore performing the maintenance work at a lower cost and usually at a better quality than when maintenance work is performed on an emergency or unscheduled basis.

In this text, preventive maintenance refers to those repairs and rebuilds that are scheduled based upon the mean time between failures (MTBF) and mean forced outage time (MFOT). Both of these parameters will be discussed later. Preventive maintenance also includes those inspections and regular maintenance activities, such as lubrication, cleaning of lines, and changing of filters, that are planned in order to prevent sudden failure of equipment and to help assure equipment is operating in a satisfactory manner.

However, in the vast majority of equipment that involves mechanical, hydraulic, pneumatic, and electrical mechanisms, it is important that certain maintenance operations take place in advance of failure in order to avoid breakdowns and extend equipment life.

Predictive maintenance utilizes a combination of cost-effective tools to obtain operating conditions of critical plant equipment, and, based on these data, maintenance is performed prior to the actual time that failure will occur. Thus in predictive maintenance, direct monitoring of critical equipment takes place so that *actual* times to failure are predicted, rather than expected time to failure based upon MTBF.

Predictive maintenance will anticipate when repair will need to take place and plans can be made for this effort. Thus, all critical equipment should be subjected to a predictive maintenance program.

There are types of equipment (principally electronic) where it is often cost-effective to perform little if any maintenance until breakdown occurs. In such cases, it is usually economical to wait until failure occurs and then replace that panel or portion of the assembly responsible for the failure. In this type of equipment, when a failure occurs there is usually a standby unit that can take over while the down equipment is scheduled for repair.

For example, in the typical disk-system desk computer, the only preventive maintenance that is cost-effective to perform is the regular (usually once a month) cleaning of the disk drive heads.

Before beginning a discussion of the application of preventive and predictive maintenance to any mechanical, hydraulic, or electrical equipment, it is necessary to have an understanding of the types of equipment failure that can occur during service. Equipment failure may be classified under four cases.

1. *Infant mortality.* These are early failures due to faulty material or faulty processing at the supplier's plant. Most of us are familiar with infant mortality when we purchase anything composed of many parts. For example, those early failures before an automobile is broken in are familiar infant mortality failures. Sound quality control and inspection procedures will minimize infant mortality.

2. *Chance failures.* Chance failures are often referred to as random failures or constant hazard failures. These are failures that take place during the product's life at a normal rate. Chance failures should not be mistaken for wearout failures. For example, automobile tires will pick up a nail at a chance rate. The 2000-plus part modern automobile has an average part failure rate of approximately 1.3% per thousand hours of service.

3. *Abuse or misuse failures.* Misuse or abuse failures are those failures where the equipment is used beyond the intended purpose. Using a piece of equipment under loads beyond its design is a typical example.

4. *Wearout failures.* These failures are those due to aging, fatigue, corrosion, and so on. These failures are characteristic of the equipment failures after long service. Wearout failures are usually progressively more frequent until the facility capability is retired because of inefficiency. Wearout can usually be postponed by proper preventive maintenance.

A theoretical relationship between the magnitude of failure and time for each of these four causes of failure is shown in Fig. 8.1. A well-conceived preventive maintenance program (predictive maintenance plus planned maintenance) can change the shape of this theoretical curve to one that appears similar to that shown in Fig. 8.2.

Not all equipment will have these four causes of failure. A quality supplier that performs in-process and final tests, including life tests and a burn-in period, will make shipments relatively free of infant mortality failures. Also, properly trained production and maintenance personnel will minimize equipment failures due to abuse or misuse. Thus, the shape of the failure rate-versus-time curve can be nearly a straight line for some components.

The entire philosophy of reliability-centered maintenance is based upon the premise that only a fraction of components conform to the traditional bathtub curve. This line of thinking was confirmed by United Airlines when they developed age-reliability patterns for the nonstructural components in their fleet. The results of this analysis are shown in Fig. 8.3.

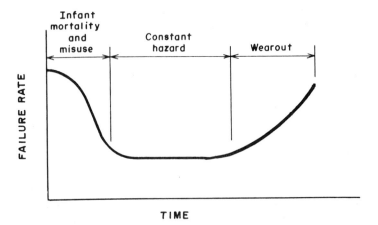

Figure 8.1. Mortality curve based on Robert Lusser's concepts.

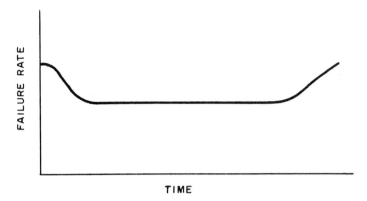

Figure 8.2 Mortality curve on a product designed and built under well-conceived quality control and reliability.

It is interesting to note that only 4% of the components studied conformed to the traditional bathtub concept, and only 6% showed a wearout period during the projected useful life of the aircraft fleet. Thus, it became apparent that a preventive maintenance system based upon the bathtub curve concept for all components would result in performing much maintenance that really is not needed. If sufficient historical data could be collected so that all components of a plant being maintained could be slotted in the appropriate failure rate-versus-time curve, then a sound preventive maintenance plan for each of the six relationships (Fig. 8.3) could be developed.

Developers of reliability-centered maintenance have established four objectives:

1. Preservation of the system function
2. Ability to analyze all functional failures
3. Prioritizing of functional failures
4. Introduction of applicable and effective preventive maintenance

Certainly it is wise to accumulate historical data so that the most appropriate preventive maintenance effort is introduced. Too often unneeded maintenance is performed that can increase the failure rate of a piece of equipment in the immediate future. A complete rebuilding of hardware can put the rebuilt equipment in the infant mortality area if the equipment falls into that class (72% of the time based upon Fig. 8.3). Also, equipment may be replaced because of fear of entering the wearout portion of the curve. Yet, in only 6% of the items in the United Airlines study was there a pronounced wearout

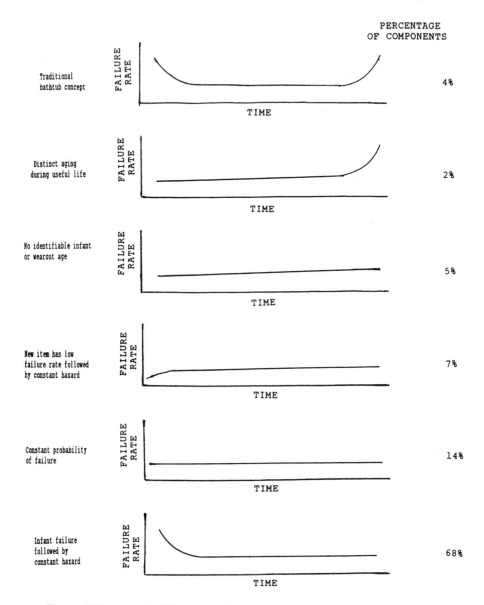

Figure 8.3 Age–reliability patterns for nonstructural aircraft equipment. (Redrawn from data from United Airlines.)

period. Many readers will acknowledge that they have traded an automobile once it has reached 50,000 miles. But this same automobile may well be driven 150,000 miles and still not have reached a pronounced wearout period.

In this volume we point out that most failures are preceded by symptoms that can be identified and measured and can be analyzed so that cost-effective predictive maintenance can be utilized.

TOTAL PRODUCTIVE MAINTENANCE

The term total productive maintenance (TPM) was first coined in Japan for a procedure in which all personnel in the organization are concerned with effective preventive maintenance. The concepts and goals of TPM are congruent with what is being advocated throughout this text. We want to use equipment throughout an organization at its maximum effectiveness. This begins with the thorough training of the operator and the broadening of the operator's responsibilities to include regular preventive maintenance functions as well as production work. Thus, operators will become thoroughly familiar with all production facilities that they operate and will care for them much as one would care for a personal car or home.

At regular planned intervals, preventive maintenance (PM) inspections will be made by the operator in concert with a specialized technician from the maintenance department, who will review those features called for under the PM planning guide. Routine PM (lubrication, oil and filter changes, cleaning, etc.) is not performed at this time since these operations take place on a regular basis by the production operator—not the maintenance department. Thus, under TPM the operator has complete autonomy in connection with all regular routine maintenance. The maintenance department can be thought of as a group of specialists who care for the equipment when unplanned failures occur and do what they can to prevent unplanned failures from taking place.

The Japanese Institute of Plant Maintenance (JIPM) identified the following features to illustrate the concepts of TPM:

1. TPM aims to change corporate culture in order to maximize overall effectiveness of production systems.
2. TPM establishes a thorough system to prevent all kinds of losses (zero accidents, zero quality defects, zero breakdowns) based on actual equipment and workplace.
3. TPM is implemented not only by production-related departments, but also by all other departments, such as development, sales, and administration.

4. TPM involves every single employee, from top management to a worker on the floor.
5. TPM attains "zero losses" through overlapping autonomous small-group activities.

SAFETY FACTOR

Before endeavoring to forecast the approximate failure date of any equipment, one should understand what the safety factor of the equipment is. When you purchase new equipment today, you are frequently given a safety factor which can be defined as the average design parameter (strength for example) minus the maximum operating parameter (stress for example) divided by the standard deviation of the design parameter. For example, let us assume a facility was designed with a given cast iron component having a mean strength of 40,000 psi and a standard deviation of 6666 psi. Now let us assume in use it is subject to a stress environment with a mean load of 17,500 psi and a standard deviation of 5840 psi. Based upon a normal distribution this strength–stress distribution would appear as shown in Fig. 8.4. In this example, based on three standard deviations, the equipment would have a safety factor (S) of

$$S = \frac{40,000 - (3 \times 5840 + 17,500)}{6666 = 0.75}$$

It is apparent that this facility with its relatively low safety factor has a rather large possible failure zone and the probability of a chance failure is significant.

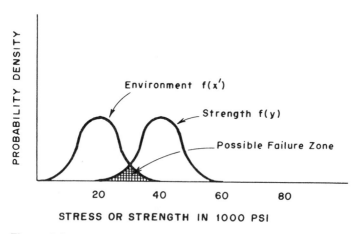

Figure 8.4 Typical product with possible failure zone shown.

It would be desirable to either reduce the environmental stress on this equipment or increase the strength of the cast iron component by replacing it with a component having a higher strength. Let us assume this component was replaced with an alloy steel component having a mean strength of approximately 60,000 psi and about the same standare deviation as the cast iron part. This change has eliminated the possible failure zone as shown in Fig. 8.5. Now the safety factor would be

$$S = \frac{60,000 - (3 \times 5,840 + 17,500)}{6,666} = 3.75$$

Another alternative would be to use a cast iron component with the same mean strength but with superior quality by reduction of its strength standard deviation (dispersion), thus minimizing the possible chance failure zone as shown in Fig. 8.6.

ANTICIPATED LIFE

In addition to specifying a safety factor in order to caution customers not to overload equipment, suppliers will frequently estimate the anticipated life of the equipment or its average use time until wearout. The majority of constant hazard failures follow a distribution that approximates the normal probability distribution with a mean failure time, T, and a standard deviation, s. Figure 8.7 illustrates this distribution and the resulting reliability which shows the

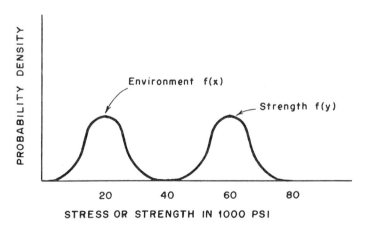

Figure 8.5 Removal of the possible failure zone shown in Fig. 8.4 by increasing the material strength.

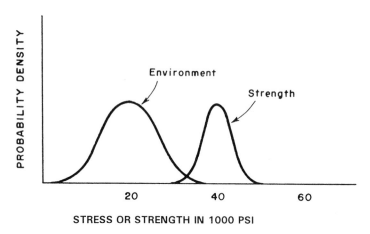

Figure 8.6 The possible failure zone can be reduced or removed by narrowing the tolerance of the material.

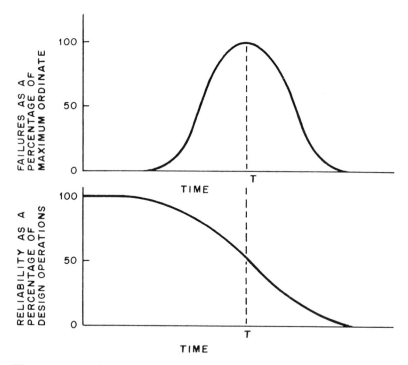

Figure 8.7 Typical wearout failure with mean failure time T.

expected percentages of a given product that are still operating after T hr of operation.

The reader should note that when constant hazard rate failures take place, the failures when plotted as a percentage of the maximum ordinate will plot as an exponential curve.

OBJECTIVES OF PREVENTIVE MAINTENANCE

The principal objectives of preventive maintenance include

1. To minimize the number of breakdowns on critical equipment
2. To reduce the loss of production that occurs when equipment failures take place
3. To increase the productive life of all capital equipment
4. To acquire meaningful data relative to the history of all capital equipment so that sound decisions as to repair, overhaul, or replacement can be made so as to maximize the return on capital investment
5. To permit better planning and scheduling of required maintenance work
6. To promote better safety and health of the work force

CHARACTERISTICS OF A PLANT IN NEED OF A PREVENTIVE MAINTENANCE PROGRAM

Any industry today that does not have a working preventive maintenance program is undoubtedly inefficient. The indications of a poorly conceived program or the lack of a program are

1. Low machine and equipment utilization because of unscheduled work stoppages due to breakdowns
2. High wait or idle time of operators because of machine or equipment downtime
3. High scrap and rejects due to unreliability of equipment
4. Increase in the cost of repairs of equipment because of neglect in regular lubrication, inspection, and replacement of worn parts
5. Decrease in the expected productive life of capital equipment because of lack of adequate maintenance

With the introduction of a soundly conceived preventive maintenance program, there will be a reversal of the five points cited above and a progressive improvement in the return on investment for facilities and equipment. Preventive maintenance, properly instituted and administered, is a cost savings program.

A good test in order to determine the adequacy of a preventive maintenance system is to respond to the following 17 questions. A score of less than 55 is an indication that the preventive maintenance program needs improvement. A score of less than 20 indicates that no real preventive maintenance program exists.

Preventive Maintenance Questionnaire

1. Is there an organized and formal preventive maintenance program in operation?
   Yes   5 _____
   Informal program only   1–4 _____
   No   0 _____

2. Is overall responsibility for preventive maintenance assigned to one person?
   Yes   5 _____
   No   0 _____

3. Are inspectors assigned full-time to inspection duties?
   Yes   5 _____
   Inspections considered part-time activity   1–4 _____
   No   0 _____

4. Are inspection routes established and scheduled based on work measurement techniques?
   Yes   5 _____
   Sometimes   1–4 _____
   Never   0 _____

5. Are check sheets controlled so that 100% compliance is assured?
   Yes, full control   5 _____
   Spot check only   1–4 _____
   No   0 _____

6. Are inspection reports spot checked by foreman to determine the reliability of the report?
   Yes, well-disciplined checking procedure   5 _____
   Occasionally   1–4 _____

7. Are plant and building assets inspected periodically as part of the formal inspection program?
   Yes   5 _____
   Partially   1–4 _____
   No   0 _____

8. Are potential problems (imminent damage or breakdown) promptly reported as a result of PM inspections?
    Yes, breakdowns definitely avoided   5 _____
    Sometimes   1–4 _____
    No   0 _____

9. Is lubrication accomplished through the scheduled utilization of check sheets?
    Yes   5 _____
    Sometimes   1–4 _____
    No   0 _____

10. Are lubrication needs reviewed periodically to reduce the number of different lubricants that are required?
    Yes   5 _____
    No   0 _____

11. Are lubrication routes established and scheduled based on time and methods studies?
    Yes   5 _____
    Sometimes   1–4 _____
    No   0 _____

12. Are meaningful downtime reports submitted to maintenance management?
    Yes   5 _____
    Needs improvement   1–4 _____
    No   0 _____

13. Is the downtime trend recorded and reported?
    Yes, in meaningful detail   5 _____
    Needs improvement   1–4 _____
    No   0 _____

14. What is the percentage of downtime for maintenance reasons?
    8% or less   5 _____
    More than 8%   1–4 _____
    Not known or used for control purposes   0 _____

15. Are breakdown reports analyzed for the detection of failure patterns that can be corrected by adjustments in the preventive maintenance program?
    Yes, definitely used   5 _____
    Occasionally   1–4 _____
    No   0 _____

16. Is data processing utilized for scheduling and reporting preventive maintenance inspections and lubrication?
    Yes, effective system in use   5 _____
    Partial use   1–4 _____
    No   0 _____

17. Are the various types of preventive maintenance work identified in the cost-reporting system permitting routine analysis of preventive maintenance as a separate category of expense?

Yes   5 ____
Partial identification   1–4 ____
No   0 ____

MACHINE AND EQUIPMENT RECORDS

We have learned that preventive maintenance (PM) is synonymous with maintenance that is performed on a scheduled basis. Thus if a preventive maintenance system is to be effective, it should include regular inspection, scheduling, issuing work orders, maintaining inventory levels of stores and stock, assigning standard times, checking performance, assembling costs, and analyzing results.

The first step in introducing a PM system is to make a record of all equipment including historical information as to the repairs and trouble encountered on each piece of equipment. In Chapter 2, we learned of the importance of maintaining a history of all production facilities. Equipment history information can be readily acquired from a review of closed work orders and equipment records. If equipment records do not exist, they should be developed to include the following information: department where facility is located, plant asset number, description of machine or facility, MTBF, frequency of PM, frequency of inspection, total time by craft for inspection, model of machine, manufacturer's serial number, manufacturer's warranty, number of failures and time required for repair, facility priority code, and life expectancy code (see Fig. 2.14).

With modern data processing equipment, regular reports relative to repair costs, downtime, lost production time, and so on, can be generated regularly by facility. In small and medium-sized plants that are not utilizing data processing facilities, a record of repairs can be maintained on a visible card file. Figure 2.14 illustrates such a record. The front of this card provides details relative to the equipment characteristics, and the reverse side provides a record of the maintenance and repair performed on the facility. When the repair records utilize all the space on the reverse side of the card, a second card is used where both sides are designed to record maintenance and repair data.

The machine and equipment records cards should provide a priority number to be associated with the facility. This priority will be helpful in the scheduling of inspections and repairs involving the equipment. Usually four priority levels will satisfactorily identify the relative importance of each facility from the standpoint of minimizing its downtime or unoperable condition. Top priority is usually coded 01 for such items as fire alarm and safety facilities and 04 assigned to those items of equipment such as air conditioning in offices where a short shutdown of service will not have a serious impact on the welfare of the personnel or the operation of the facility. Thus these codes can generally be used:

01 Emergency
02 Rush
03 Important
04 Routine

Another item of information that should be provided on the equipment record is an estimate of its life expectancy. This typically will range from one to thirty years.

With a complete record of all equipment, including recommended inspection and lubrication schedules, it is now possible to initiate work orders and schedule this phase of PM. The work order procedure was discussed in Chapter 2 and scheduling in Chapter 7.

PM INSPECTION

What mechanical, pneumatic, hydraulic, electrical, and electronic inspection should be performed and at what intervals are important components of the PM program. To determine what should be inspected, the maintenance analyst will utilize three sources of information. First, the manufacturer of the equipment will be able to provide helpful information as to recommended maintenance, lubrication, and overhaul schedules. This information is most useful in determining what should be inspected. Second, a review of the breakdowns and repairs that have taken place in the past will provide a guide as to the nature of recommended inspections. Finally, consultation with the operators, craftsmen, and line supervisors who are closely associated with the operation and maintenance of equipment can provide helpful suggestions.

In the initial stages of a PM program, there is a tendency to spend too much time in connection with the inspection procedure. Not only are unnecessary inspections made, but the method utilized in the inspection is often crude and does not conform to sound principles of work simplification and utilization of the correct diagnostic tools and procedures (see Chapter 9).

Preventive maintenance on one facility can be entirely different from that on another machine. Even equipment produced by the same builder will have different areas that are subject to wear. In machine work this will depend on the particular jobs being run and the type of tooling employed.

A representative PM inspection to be performed on 4- and 6-spindle automatic screw machines may include:

1. Remove and clean collets and tubes.
2. Check chucking fingers, pins, and cam rollers.
3. Check the wipers and gibs on all slides.
4. Check both cross and tool slide cams.
5. Check brake. Adjust if necessary.
6. Check clutches. Adjust if necessary.
7. Check filters. Clean if necessary.
8. Check lubrication oil lines. Clean if necessary.
9. Check for loose bolts. Tighten if necessary.

Figure 8.8 illustrates a representative preventive maintenance check card. A card of this nature can be attached to each facility and when the equipment is serviced, the maintenance inspector will record what was done.

After determining what is to be inspected, the ideal methods for performing these inspections are planned along with the time estimates, including the standard minutes to perform the inspection. For example, in checking temperature control, the type of service is considered and analyzed in order to determine the best instrument to do the job. Before selecting the exact instrument, the following factors are considered:

1. Degree of accuracy and sensitivity needed
2. Range of temperature involved
3. Proximity of instrument from point of measurement
4. Safety equipment that may be needed

The reader should understand that the period between inspection of different components of a piece of capital equipment will vary considerably. For example, it may be desirable to check a transformer for excessive heat every 3 mo. but check the drive pump motors for bearing conditions, end play of shaft, and alignment to component parts driven by the same motor annually.

The total inspections performed at scheduled intervals on equipment are recorded on a check sheet. The check sheet will identify what is to be done at periodic intervals. Thus it may provide information as to what is to be checked every 30 days, what additional checks are to be made every 3 mo, and those further checks to be performed annually. The check sheet provides information for issuance of the PM work order.

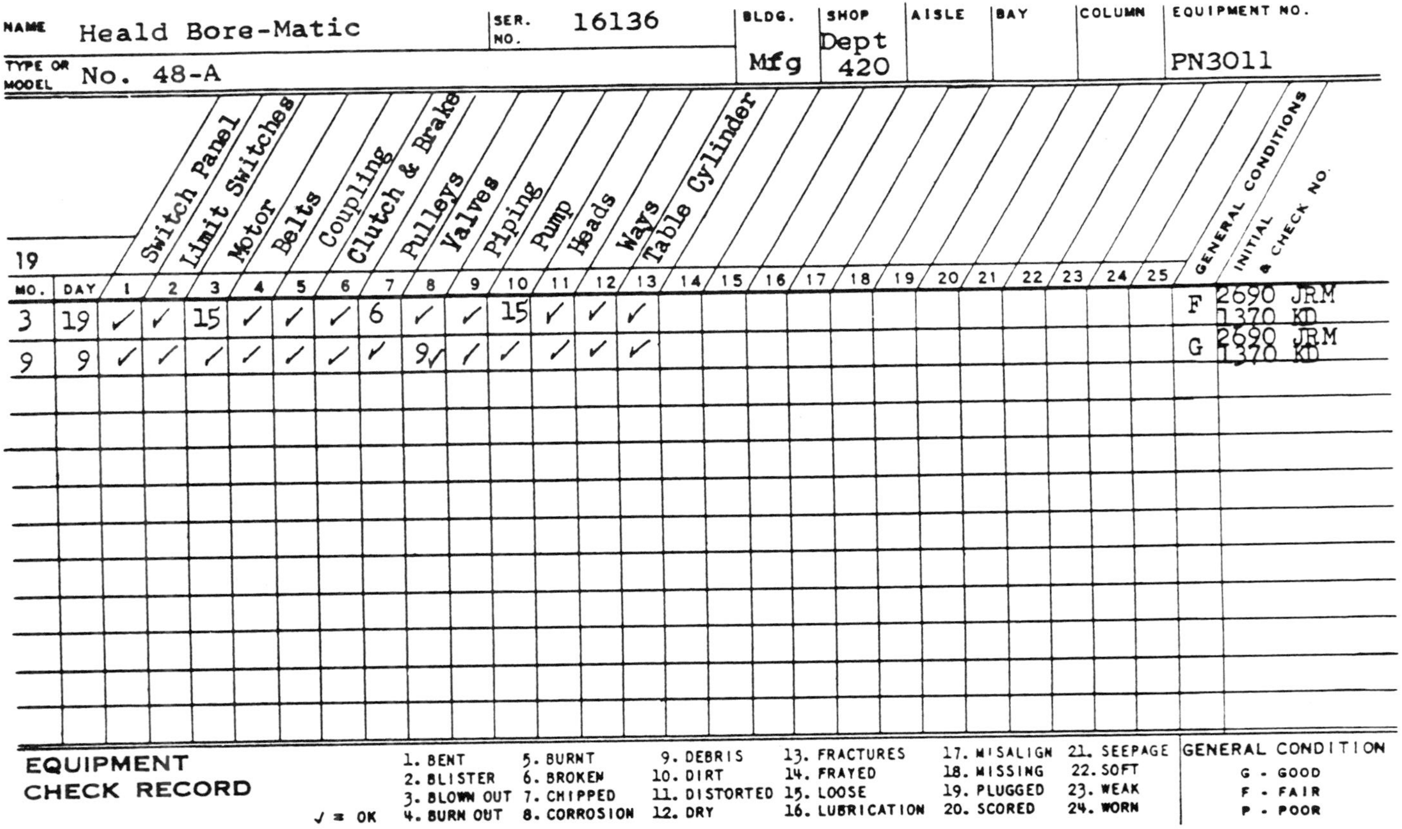

NAME Heald Bore-Matic   **SER. NO.** 16136   **BLDG.** Mfg   **SHOP** Dept 420   **AISLE**   **BAY**   **COLUMN**   **EQUIPMENT NO.** PN3011

TYPE OR MODEL No. 48-A

19	MO.	DAY	1 Switch Panel	2 Limit Switches	3 Motor	4 Belts	5 Coupling	6 Clutch & Brake	7 Pulleys	8 Valves	9 Piping	10 Pump	11 Heads	12 Ways	13 Table Cylinder	14	15	16	17	18	19	20	21	22	23	24	25 GENERAL CONDITIONS	GENERAL CONDITION	INITIAL & CHECK NO.
	3	19	✓	✓	15	✓	✓	6	6	✓	✓	15	✓	✓	✓													F	2690 JRM / 7370 KD
	9	9	✓	✓	✓	✓	✓	✓	✓	16	✓	✓	✓	✓	✓													G	2690 JRM / 7370 KD

✓ = OK

1. BENT	5. BURNT	9. DEBRIS	13. FRACTURES	17. MISALIGN	21. SEEPAGE
2. BLISTER	6. BROKEN	10. DIRT	14. FRAYED	18. MISSING	22. SOFT
3. BLOWN OUT	7. CHIPPED	11. DISTORTED	15. LOOSE	19. PLUGGED	23. WEAK
4. BURN OUT	8. CORROSION	12. DRY	16. LUBRICATION	20. SCORED	24. WORN

GENERAL CONDITION
G - GOOD
F - FAIR
P - POOR

EQUIPMENT CHECK RECORD

Figure 8.8 Preventive maintenance check card.

Inspection Frequency

The frequency of inspection is a function of the type of equipment, its age and condition, the utilization, the working environment, and the consequences of an unscheduled shutdown due to failure. In general, equipment is inspected either weekly, monthly, quarterly, semiannually, or annually, as a PM function. Where more regular inspection is required, such as once every shift or once every 24 hr of service, the inspection is incorporated in the work standard and performed by the production operator. For example, an operator using a "S" type recorder would as part of his setup check visually each day to ascertain that his recorder is functioning properly. He would check the condition of the battery, chart, and recording pen. He would change the chart as required. However, every 30 days the PM department may check the signal and panel lights for proper operation.

Once inspection frequencies are determined and implementation takes place, there should be a program of follow-up to assure planned intervals between inspections are of the correct duration. Where downtime is minimal, it may be cost-effective to extend the period between inspections. Similarly, where downtime is occurring more often than anticipated, it may be desirable to shorten the time between inspections. After the inspection frequency of a facility is estimated, this information should be identified by a code number and posted to the machine or facility record card. For example, the following might be used.

Code no.	Frequency
01	Daily
02	Weekly
03	Biweekly
04	Monthly
05	Every three months
06	Semiannually
07	Annually

Determining the Inspection Route and Equipment

Considerable travel time can be saved if the PM inspector utilizes standard packages of instrumentation and hardware developed for conducting inspections on all the equipment that will be on his route for at least one-half of his working day. The standard tool kit should be identified at the time the ideal method is established. For example, a kit may contain open-end and box wrenches, screwdrivers, a speed indicator, torque wrench, a multitester for taking ac voltage measurements, resistance measurements, and decibel measurements, an infrared meter to check for loose connections and overheating, a feeler gauge for measuring small clearances, and a dial indicator setup for checking concentricity. Another kit for making different inspections would contain an entirely different set of hardware and instrumentations. Utilizing such planned tool kits will not only save time in checking out tools and returning them to the maintenance tool crib, but also will save the time required to make additional trips to the tool crib to obtain tools forgotten or overlooked when making subsequent inspections after the first inspection.

At the time the standard kit of tools is checked out, a route sheet should be provided that gives the pattern of travel that results in the shortest travel distance by the inspector with due consideration for the production work that has been scheduled to the equipment. Thus the travel route (established by the planning function) will not only take into consideration the physical location of the equipment being inspected but also its availability for inspection without causing a cessation in productivity.

The travel time needed to move the craftsmen to and from the job site can be determined with the assistance of a travel chart. Figure 8.9 illustrates the travel time in decimal hours from and between a central repair shop and six facility sites (four buildings and two areas). Charts similar to the one shown should be developed to aid in establishing standard travel times between buildings and areas.

PLANNED MAINTENANCE

A large percentage of planned maintenance is a result of scheduled inspection. However, there are PM operations such as lubrication and overhauls that are scheduled at regular intervals in view of experience and the recommendations from the producers of the equipment. The total amount of scheduled maintenance, which should be the vast majority of all maintenance work performed (typically about 85%), then is made up of those work orders initiated because of PM inspections and those PM work orders initiated in order to help assure that the production equipment stays in working order for extended periods.

Lubrication

An effective lubrication program is one of the key ingredients of a sound PM system. Lubrication is performed for several purposes. The principal reasons include the reduction of friction between rolling and sliding contacts and the

	Shop	Bldg. 1	Bldg. 2	Bldg. 3	Bldg. 4	Area A	Area B
Shop		0.1	0.2	0.2	0.3	0.4	0.5
Bldg. 1	0.1		0.2	0.3	0.4	0.5	0.6
Bldg. 2	0.3	0.2		0.5	0.3	0.8	0.3
Bldg. 3	0.2	0.3	0.5		0.2	0.7	0.4
Bldg. 4	0.3	0.4	0.3	0.2		0.7	0.2
Area A	0.4	0.5	0.8	0.7	0.7		0.9
Area B	0.5	0.6	0.3	0.4	0.2	0.9	

Figure 8.9 Area travel times.

minimization of wear between mating surfaces. For example, as gears mesh and make contact the action consists first of sliding, then rolling, and then sliding again. Sliding predominates at the initial point of contact but then diminishes. When teeth are fully engaged and make contact at the pitch line, they no longer slide, but roll over each other. Then, as they disengage sliding action begins again and increases (see Fig. 8.10). This sliding action forms a lubricating wedge which may totally separate the mating surfaces, helping to support the load and reducing friction and surface wear. Lubrication is performed also to provide cooling, to prevent corrosion, and to act as a protective coating to seal out moisture and dirt.

It is necessary to identify both the proper lubricant and the lubrication frequency for all the equipment in the plant. For each piece of equipment, a master lubrication card should be prepared. This card includes the following information: machine number, machine name, machine location, lubrication

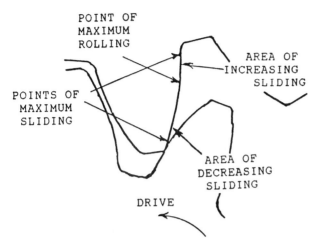

Figure 8.10   Meshing of gear teeth includes sliding and rolling.

points, type of lubricant at each lubrication point, lubrication frequency at each lubrication point, check-off space to show that lubrication is completed, and a space to identify any observed irregularity at each lubrication point.

The various types of machines typical of those used in diversified industry may require different types of oil or grease. It is not unusual for a company to stock as many as 30 different categories of lubricants in addition to cutting oils. For example, electric motor and generator bearings are usually either oil- or grease-lubricated. If oil-lubricated the viscosity recommended is in the range of 150 to 360 SUS at a 100°F (see Fig. 8.11). The oil may be a straight mineral oil or may include an antirust or antioxidant inhibitor. Today many motors have their bearings grease-lubricated by the manufacturer, which will last as long as the expected life of the motor. Those motors that require periodic lubrication with grease usually use either a No. 1 or 2 grade of either sodium or lithium grease.

In the interest of economy, the inventory of lubes maintained should be studied to see if keeping fewer on hand may not be more cost-effective. Some of the lubricants produced in recent years are highly versatile so that lubrication requirements may be simplified. For example, some oils today not only provide good stability at high temperatures and wet conditions, but also offer good protection against wear. It is also desirable to have a cross-reference chart so that different producers' lubrication numbers can be correlated.

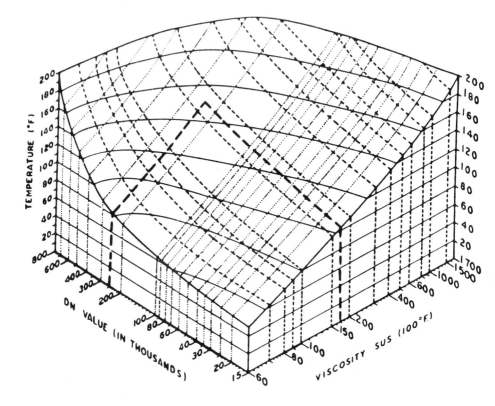

Procedure:

1. Determine the DN value—Multiply the bore diameter of the bearing (in mm) by the speed of the shaft (in rpm).

2. Select the proper temperature—The operating temperature of the bearing may run several degrees higher than the ambient temperature depending upon the application. The temperature scale shown reflects the operating temperature of the bearing.

3. Enter the DN value on the DN scale (250,000 on chart).

4. Follow or parallel the "dotted" line to the point where it intersects the selected "solid" temperature line (150° F on chart).

5. At this point follow or parallel the nearest "dashed" line downward and to the right to the viscosity scale.

6. Read off the approximate viscosity value—expressed in Saybolt Universal Seconds (SUS) at 100° F.

Figure 8.11   Oil viscosity selection chart. (Courtesy of U.S. Army Material Command. *Engineering Design Handbook.* AMCP, 1972.)

Lube Oil Analysis

A lube oil analysis program should be included in a modern engineering maintenance effort. Regular lube oil analysis will help to assure more reliable equipment performance and reduce the amount of unscheduled downtime. Lube oil analysis can provide information on the condition of a facility and the condition of the lubricating oil. Oil analysis tests will determine if fuel or soot have contaminated the lube oil, whether coolant has entered the oil stream, and whether dirt or metal dust is entering the engine. If any of these contaminants are present, the engine will not operate as effectively as it should. For example, water in lube oil can result in corrosion of internal engine parts. If coolant contains glycol, the engine could seize as the lubricating quality of the oil becomes severely retarded.

Analysis of metals carried by the lubricating oil can identify points of engine wear. For example, high levels of copper would indicate worn bearings, excess chromium might indicate high wear on chromium-plated piston rings, and iron would indicate wear from gears, shafts, and so on. It is evident that with oil analysis, problems can be identified and corrected before necessitating a complete overhaul.

Oil analysis also identifies if the correct oil viscosity and correct additives are being used. Oil that has been developed for elevated temperatures will not perform as well at lower temperatures.

Methods of Application

The correct method of lubrication is an important factor in obtaining good performance for most moving parts. A variety of lubrication systems are characteristic of machinery and equipment being utilized by modern industry. Grease cups, gravity-fed oilers, bottle oilers, air-mist lubricators, and constant-level lubricators all will be encountered by the maintenance mechanic who is servicing the equipment in the typical plant (see Figs. 8.12 through 8.15).

Where circulating and bath contact systems are used, as in crank cases, there usually is a filtration system for the removal of foreign matter picked up in the circulating lubricant. This filter, if cartridge or bulk type, will need to be cleaned or replaced at regular intervals (see Fig. 8.16). Thus, the service functions related to lubrication include: changing coolant or lubricant, filtering coolant or lubricant, flushing troublesome system, changing hydraulic system, and so on. It is not unusual to have as many as one hundred or more different routine equipment service functions, totaling thousands of operations in a medium-sized to large plant.

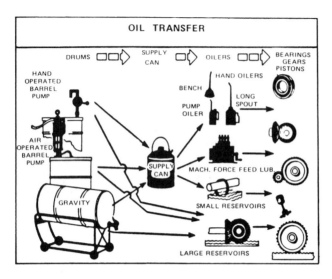

Figure 8.12 Oil transfer. (Courtesy of U.S. Steel Corporation.)

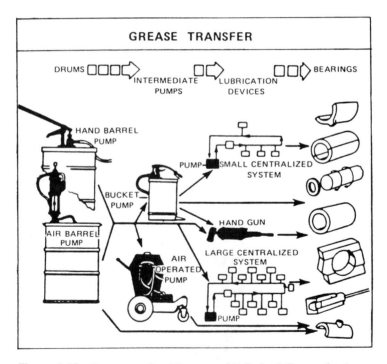

Figure 8.13 Grease transfer. (Courtesy of U.S. Steel Corporation.)

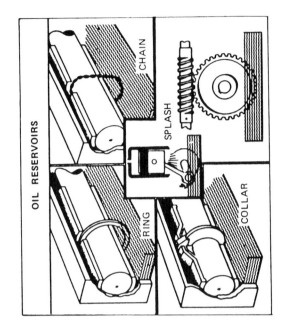

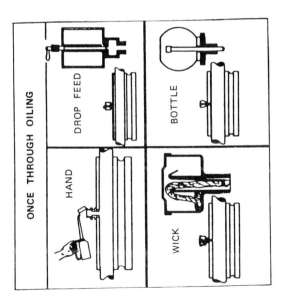

Figure 8.14 Once-through oiling and oil reservoirs. (Courtesy of U.S. Steel Corporation.)

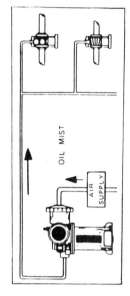

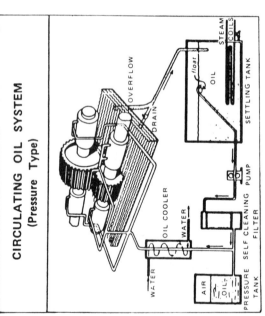

Figure 8.15 Circulating oil system (pressure type). (Courtesy of U.S. Steel Corporation.)

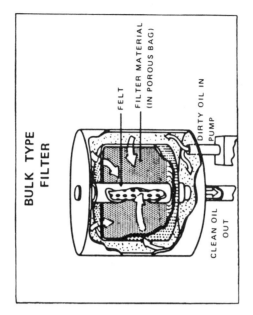

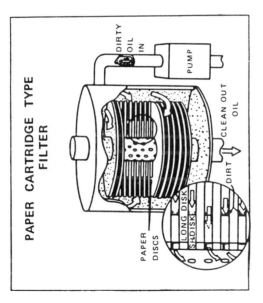

Figure 8.16a Paper cartridge type filter and bulk type filter. (Courtesy of U.S. Steel Corporation.)

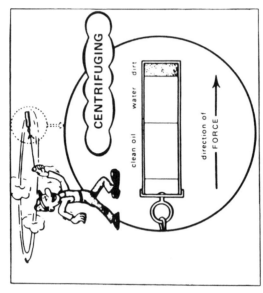

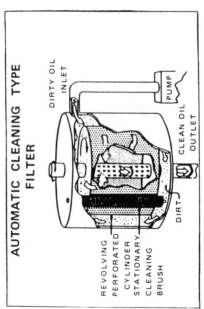

Figure 8.16b Automatic cleaning type filter and centrifuging. (Courtesy of U.S. Steel Corporation.)

In many instances, automatic lubricators can be installed to update the manual lubricators that were supplied with the original equipment, thus reducing the frequency of application periods and conserving lubrication personnel time. The opportunities for methods improvement should not be overlooked in this facet of maintenance work. Often, special devices can be used to permit quick connection of couplings to portable filters; improved grease-gun loading may be developed; the utilization of high-volume pumps in order to reduce the time in changing reservoirs of lubricants can perhaps be introduced; and reducing loss of lubricants by leakage might be possible. For example, Fig. 8.17 illustrates a typical arrangement of splash-lubricated drive gears which are enclosed in a two-piece steel casing. The oil circulates through a system consisting of the casing, a tank, pump, and piping. The shaft extends through an opening in the side of the casing where a seal or packing prevents the escape of oil around the shaft. Leakage may occur at the pipe joint, at the seal, or at cracks that develop in the casing. The following steps should be taken in order to minimize leakage:

1. Use the proper seal materials and joint types for the lubrication fluid used.
2. Make sure all connections are accessible for inspection and maintenance. Use as few connections as possible.
3. Be sure that valves and tubing are protected from possible damage by factory traffic.
4. Be sure that work orders for performing lubrication state that a check for leaks be made.

In addition to the problem of preventing oil leakage, there is the problem of contamination of the oil by foreign substances or solutions. Typical contaminants include water and coolant solutions that are used in large quantities in the machine tool and steel industries. Often these liquids enter oil systems through loose oil seals around shafts and bearings and through small openings in gear casings. In many plants, production equipment is exposed to dust and fine solids such as cement, coal dust, mill scale, raw materials, and finished products. All of these contaminants may find their way into the oil. When wear occurs on gears and bearings, the fine particles of metal are also carried in the oil as contamination. Oil which has been in service a long time will deteriorate to some extent by combining with the oxygen in the air.

Solid contaminants are usually abrasive and will cause excessive wear on moving parts. They may also plug oil passages and cause bearing failure because of lack of lubricant. The presence of large amounts of solid contaminants usually can be determined by a general inspection of the interior of gear

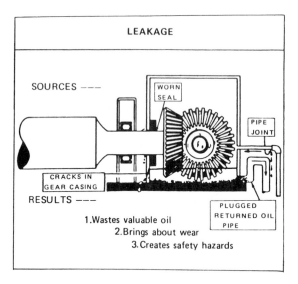

Figure 8.17   Leakage. (Courtesy of U.S. Steel Corporation.)

cases and reservoirs. However, the best method of determining the condition
of oil is a laboratory test, which will identify both the type and amount of
contaminants present. It is especially important to have laboratory tests made
periodically on the lubricant of large circulating systems in order to be sure
that the oil is being kept in the best possible condition for good lubrication.
Inexpensive testing laboratories are usually available in most metropolitan and
industrial areas.

   Contaminants can be removed from oil by filtering, centrifuging, or settling
(see Fig. 8.15).

   The lubrication standard practice procedure (master lubrication card) should
be maintained for all machinery and equipment. This information is used for
developing the work schedule assigned to the workmen who perform the
lubrication. Table 8.1 illustrates a master lubrication card that provides standard practice procedure for a range of sizes of Bullard lathes.

   It is good practice to make a workload analysis of the lubrication function.
This analysis considers time values, lubrication frequencies, travel distances,
and methods. The assignment of the number of oilers should be based on this
analysis.

Table 8.1 Lubrication Standard Practice Procedure

Machine(s): 30-in., 36-in., and 42-in. Bullard vertical lathes

Machine location: Building 82

General Lubrication Information: Oil in the clutch case bracket reservoir is pumped through a filter to lubricate the table bearings, gear and pinion, headstock clutch and brake, and gear shift mechanism. All brackets on top of the machine and gear trains within the various heads are lubricated from their own independent reservoirs.

Lubrication Schedule (8-hr daily operation):

1. *Three times per week.* All Zerk fittings including those on slide and ways, feed screws, binder shaft, rail raising brackets, rail turret heads, rail raise heads, and side heads. Use SAE50 oil with viscosity range of 80 to 105 seconds Sayboldt at 210°F.

    Standard hours: 0.18

2. *Once per week.* Check hydraulic reservoirs and if necessary fill with 150 S.S.U. at 100°F turbine oil with rust and oxidation inhibitors.

    Standard hours: 0.15

3. *Once per month.* Lubricate cross rails and vertical slides. Grease chucks with lime base grease.

    Standard hours: 0.21

4. *Once every six months.* Lubricate motors. Change central lubricating filter cartridge.

    Standard hours: 0.08

Lubrication Fittings

All grease fittings should be easily accessible. When this is not the case, the equipment should be adapted by the addition of extension lines in order to bring the grease fitting to an accessible location. The fitting attached to the end of the line should be securely anchored in order to withstand heavy duty. To simplify lubrication throughout the equipment, it is desirable to have fittings of the same size and type. Figure 8.18 illustrates typical lubrication

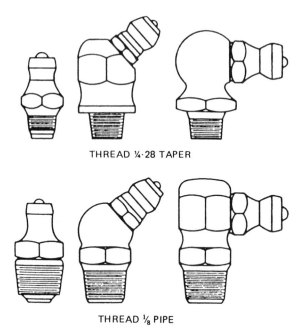

THREAD ¼·28 TAPER

THREAD ⅛ PIPE

Figure 8.18   Typical lubrication fittings.

fittings. The following positive design considerations will reduce servicing time, provide adequate lubrication, and reduce the amount of dirt introduced into the lubrication system.

1. Utilize a central mechanism for applying the lubricant rather than through multiple lubrication fittings.
2. Provide lubrication fittings and reservoirs for all types of plain annular and plain self-aligning bearing installations (see Fig. 8.19).
3. Oil seals should be easy to replace.
4. Dipsticks should be provided where necessary to measure oil levels.
5. Provide magnetic chip detectors equipped with warning lights in lubricating systems wherever ferrous chips may endanger the life or operation of the equipment.

Synthetic Lubricants

The role of synthetic lubricants of the synthesized hydrocarbon type have a promising future in reducing energy consumption and consequently reducing

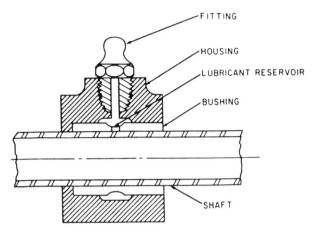

Figure 8.19 Lubrication fitting, lubricant reservoir, and lubrication grooves for plain bearing–bushing installation.

the cost of power per unit of product output. Synthetic lubricants have been designed with lower coefficients of traction than mineral oil, thus less energy is required to shear the lubricant film which must take place to allow relative motion between a moving part and a bearing surface. It has been reported that synthetic greases, when compared with conventional lithium grease, require lower power demand on start-up in view of a lower breakaway torque. The percentage savings increase significantly with lower temperatures (below 0°F).

Both synthetic oils and greases have proven to be cost effective not only by the reduction of power costs, but also through longer oil change intervals, improved component life, and less facility downtime.

A PREVENTIVE MAINTENANCE MODEL

An example of how one company, a drop-forge company, introduced a preventive maintenance program in connection with their forging hammers will illustrate the economic advantages of this type of effort.*

The company was experiencing a great deal of downtime for a variety of reasons. Each downtime was resulting in emergency-type maintenance in order to get the forging hammer back into production. The company had no records on the mean time between failures of the forging hammers nor did they have records on the average repair time for the various causes of malfunction.

*Courtesy of Dr. Kenneth Knott.

The first step in introducing a preventive maintenance program is to accumulate reasonably close estimates of the parameters for mean time between failures (MTBF) and the total time to make the repair including time to acquire needed parts and materials (MFOT). This was accomplished by choosing a small team of those most qualified and experienced to handle the estimating. This team included the foreman of the maintenance effort, the plant engineer, and the principal maintenance employee involved with these repairs.

This team first developed a listing of all those failures that took place on the forging hammers. This included

1. Pins on linkage
2. Treadle bolts
3. Adjust clamps
4. Wedge bolts
5. Gib bolts
6. Rocker arms
7. Shocks
8. Oil lines

The team went through an estimating procedure as shown in Fig. 8.20. The estimating sheets required each member of the team to make three estimates of the lost time for each of the eight failure modes. These three estimates included:

T_o = Estimated optimistic time to make the repair
T_m = Estimated average time to make the repair
T_p = Estimated pessimistic time to make the repair

Although each team member made estimates independently, the estimates of the other team members were shown at the next meeting of the group where all estimates were discussed. A second estimate by all team members of the T_o, T_m, and T_p values were made. This was expected to result in more reliable estimating since each team member would benefit from the interaction with the others. Finally, a second meeting of the team took place and a third estimate by each team member was made.

Not only were three estimates made for the MFOT, but the frequency per 80 hr of operation for each failure mode was made. With the last estimate by the team members, average values were computed for T_o, T_m, and T_p. Then, using the triangular distribution, expected times for the repair of each of the eight faults were made. These calculations were based on

$$\text{Expected time} = E(T) = \frac{T_o + T_m + T_p}{3}$$

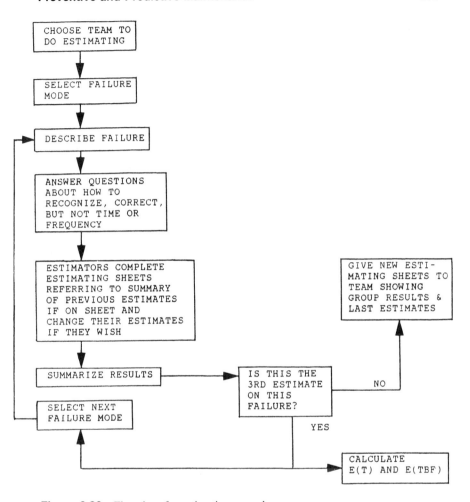

Figure 8.20 Flowchart for estimating procedure.

For example, if T_o = 1 hr and T_m = 2.4 hr and T_p = 5.5 hr, $E(T)$ would equal 2.97 hr (see Fig. 8.21). Using this procedure, expected times were estimated for the repair times for all eight faults. Estimates for the frequency of repair for each of the eight faults were computed by averaging the estimates of the team members based on their third estimate. Table 2 summarizes these estimates.

Control charts, as shown in Fig. 8.22, were then maintained for each facility. This permitted a check on all estimates being used.

FREQUENCY

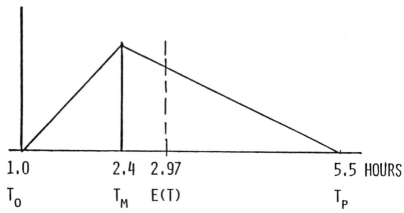

$$E(T) = \frac{T_0 + T_M + T_P}{3}$$

Figure 8.21 Triangular distribution.

Table 8.2 Summary of Estimates of Lost Time

	T_o	T_m	T_p	$E(T)$	Estimated frequency per 80 hours	Total hours
1. Pins on linkage	0.50	1	2	1.17	1	1.170
2. Treadle bolts	0.50	0.75	2	1.08	1	1.080
3. Adjust clamps	0.50	0.75	1	0.75	1	0.750
4. Wedge bolts	0.75	1	2	1.25	0.50	0.625
5. Gib bolts	0.50	0.75	1	0.75	1	0.750
6. Rocker arms	1	1.50	2	1.50	0.50	0.750
7. Shocks	0.50	0.75	1	0.75	0.17	0.125
8. Oil line	0.50	0.75	1	0.75	1	0.750
Total hours lost per 80 hours production						6.000

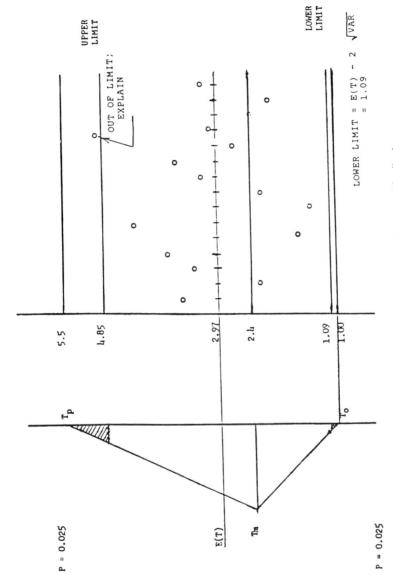

Figure 8.22 Establishing upper and lower control unit with the triangular distribution.

Finally, and most importantly from the standpoint of preventive maintenance, a decision diagram was established at each forging hammer. Every time a die change was required, the forging operator, using the decision matrix, made a decision as to whether or not to incorporate preventive maintenance for each of the eight faults while the forging hammer was down for die change. Figure 8.23 illustrates the operation of the proposed preventive maintenance system.

The installation of this preventive maintenance system resulted in a reduction of lost downtime of the forging hammers by 54% in the first 6 weeks after installation. The predicted annual savings of production workers was 58,320 man-hours in the first year after installation.

PREDICTIVE MAINTENANCE

Under predictive maintenance, the maintenance department is able to get maximum usage of capital equipment prior to any failure and resulting downtime. Preventive maintenance, based upon the statistics of MTBF and MFOT, will permit scheduling of repairs in advance of actual failure, but the time between the actual repair and the time the facility could have safely operated before a breakdown occurred could be appreciable. The whole purpose then of predictive maintenance is to maximize key facility uptime with very low probability of required unscheduled maintenance.

To do this, one must continually monitor the mechanical condition and the system efficiency to be able to predict the precise failure time with close accuracy. There are five nondestructive techniques typically used to do this monitoring. These are

1. Vibration monitoring
2. Process parameter monitoring
3. Thermography
4. Tribology
5. Visual inspection

With a well-designed predictive maintenance system in place, one can expect to eliminate catastrophic machine failures, eliminate unnecessary repairs, and make necessary repairs on a scheduled basis using the most cost effective procedures.

In the next chapter, on diagnostic techniques, considerable detail will be presented as to the nondestructive techniques that may be used to implement a positive predictive maintenance effort.

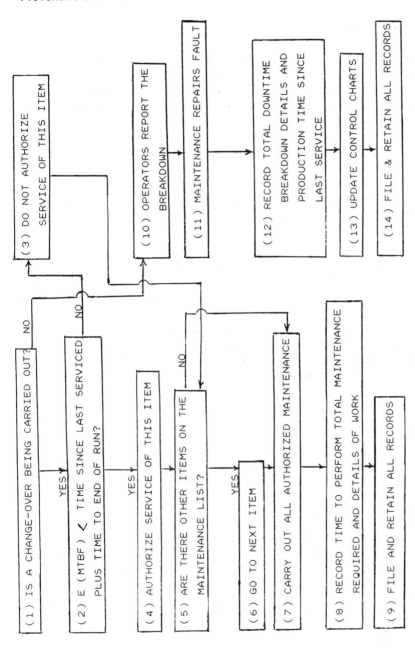

Figure 8.23 Operation of proposed preventive maintenance system.

SUMMARY

Just as preventive medicine is usually less expensive than corrective medicine, so is preventive and predictive maintenance cost-effective in comparison to breakdown maintenance. Preventive and predictive maintenance result in no facility downtime or deferrable downtime while breakdown maintenance will result in either downtime or deferrable downtime (see Fig. 8.24). The initial step in establishing a PM program is obtaining and organizing the pertinent data. The equipment to be included in the program needs to be identified, and a listing of the inspection jobs to be performed on each piece of equipment needs to be made. Equipment should incorporate a method of identification so that PM operations can be reduced or discontinued during periods of low productivity when the use of certain equipment may be reduced or discontinued.

Each inspection assignment is consecutively numbered in order to permit the best sequencing. The planning of the PM program should follow the general flow of the product or products through the production equipment.

Planned repair orders are developed from the information obtained from the PM inspections. With this collective listing of planned repairs to be done, it is a simple matter to determine job priorities and to effectively schedule the maintenance repairs. The PM inspection sheet should include the location and fault to be repaired, the date the problem was discovered, and the urgency of the repair. If no problems are discovered, the inspector will mark the sheet O.K.

There should be three types of inspections performed under the PM program: operational inspections, replacements inspections, and inspections that require facility shutdown.

Operational inspections are those of a visual or listening nature which do not require that the equipment be shut down (see Fig. 8.25). Frequently, operational inspections will utilize various diagnostic techniques, including sensing devices such as temperature gauges, stethoscopes, and vibration detectors (see Chapter 9).

Replacement inspections are those inspections of a specific component or components that have been scheduled for replacement at specific intervals. The part that was in service may be returned to the supply area where it will be inspected and serviced and repaired if necessary and returned to stores as a usable spare. In the event it is not cost-effective to repair the worn component, it will be scrapped and the balance of stores ledger appropriately posted. The replacement inspection may result in deferring the replacement of the inspected component.

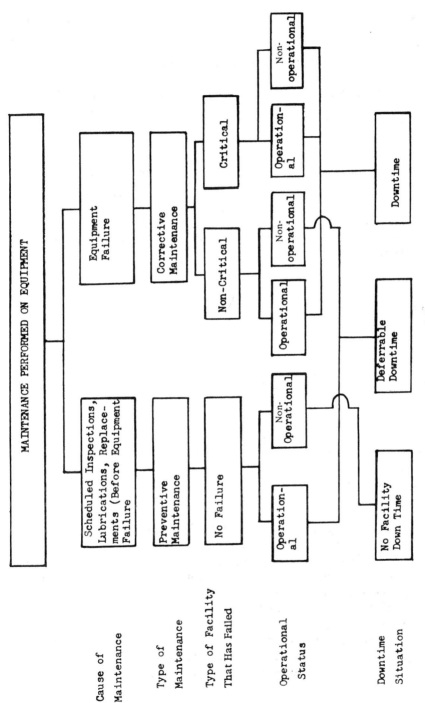

Figure 8.24 Downtime classification.

PREVENTIVE MAINTENANCE OPERATIONAL INSPECTION

ORDER NUMBER 1211   DEPARTMENT 42" & 48" Cold Tin Mills

UNIT #6 skin pass   FACILITY Feed Reel Lines

JOB INSTRUCTIONS

Check piping for damage and leakage. Inspect hoses for kinks and chafing. Examine connections for leakage. Inspect for loose or missing hangers.

REMARKS Pipe plugs missing at oil check level on both brackets. Small pool of grease at north and south ends of line.

SIGNED   *Cl. B Jones*   DATE   *12/15*

RETURN TO FOREMAN WHEN COMPLETED

Figure 8.25 Preventive maintenance operational inspection of feed reel lines in 42-in. and 48-in. cold tin mills.

    Shutdown inspections will frequently include the use of measuring devices, nondestructive test equipment, and visual inspection. In the event defects are found, the components will be scheduled for replacement. Here the production equipment will need to be shut down in order to complete the inspection. It is a good idea to maintain a log of the PM performed either on the equipment record card or else as a separate log. Figure 8.26 illustrates a boiler room preventive maintenance log.

    The personnel performing the inspections should be experienced, knowledgeable, reliable people who have had a sound background in their particular area of competence. Even with quality personnel performing the inspections,

BOILER ROOM PREVENTIVE MAINTENANCE LOG

VACUUM PUMPS:	JULY	AUG	SEPT	OCT	NOV	DEC	JAN	FEB	MAR	APR	MAY	JUNE
Check couplers												
Drain vaccum tanks												
Check control switches												
Service all valves												
CLEAN & SERVICE PUMPS EVERY 6 MTHS.												
SUMP PUMPS:												
Check float controls												
Check alarms												
GREASE & SERVICE PUMPS EVERY 6 MTHS.												
HEATERS:												
Clean coils												
Check filters												
Check motor and belts												
SERVICE MOTORS EVERY SIX MONTHS												
DATE INSPECTED:												
YOUR INITIALS												

Figure 8.26   A typical checklist for preventive maintenance of boilers.

it is important that the information provided on the PM inspection order is sufficiently detailed to assure a thorough inspection.

The scheduled inspection approach to PM will prove to be cost-effective. The benefits of PM will result in greater equipment availability at lower costs of operation.

Predictive maintenance is almost always cost-effective when used at those work centers involving high costs. Here, continual monitoring of those parameters that allow the accurate prediction of failure will permit precise scheduling of repairs without the high cost of emergency downtime. With scheduled maintenance, there will be an extended use of high cost equipment prior to maintenance.

SELECTED BIBLIOGRAPHY

Mann, Lawrence Jr. *Maintenance Management*, Lexington, MA: Heath, 1976.

Nakajima, S. *TPM Development Program*. Cambridge, MA: Productivity Press, 1989.

Newbrough, E. T., and the staff of Albert Ramond and Associates, Inc. *Effective Maintenance Management*, New York: McGraw-Hill, 1967.

Okazaki, E. *Investigation and Report on the Maintenance Technology of Manufacturing Plants in the Fiscal Year 1987*. Tokyo: Equipment Maintenance Information System, The Japanese Institute of Plant Maintenance, 1987.

Smith, Anthony M. *Reliability-Centered Maintenance*. New York: McGraw-Hill, 1993.

Society of Automotive Engineers, *Reliability, Maintainability and Supportability Guidebook*, 2nd ed. Warrendale, PA: SAE, 1992.

U.S. Army Material Command. *Engineering Design Handbook*. AMCP, 1972.

9
Diagnostic Techniques

In the previous chapter, we learned that an effective maintenance program incorporates three types of inspections: operational inspections, replacement inspections, and those inspections that necessitate an equipment shutdown. In all three of these inspection classes, there often is the need to utilize diagnostic techniques in order to predict maintenance by identifying a particular disorder and its magnitude. Just as the physician has many diagnostic techniques to assist in preventive medicine so does the predictive maintenance program necessitate the use of diagnostic methods. The techniques should be utilized regularly on all critical equipment so that imminent failure can be predicted and scheduled maintenance and repair can take place before breakdown.

In much of today's sophisticated equipment, including robot systems, there is a special need for diagnostics in order to be able to predict maintenance. In connection with robotics, typically up to one-half of the maintenance time is dedicated to diagnostics and the remaining time is divided between performing the actual repair and testing in order to verify the repair's quality. The mean time between failures (MTBF) of the modern robot is approximately 36 months under 8-hr daily operation. As has been pointed out, it often is not cost-effective to endeavor to predict maintenance in electronic controlled equipment. It usually is less expensive to wait until failure occurs and then replace that component, panel, or circuit where the failure exists.

Often early warning systems can be incorporated within the equipment. For example, brushes for any rotating equipment can be provided with an embedded sensing probe that can give either a visual or audible warning when it is time to rebrush.

The data needed in order to diagnose the condition of equipment includes overall vibration level, axial position indication, bearing temperatures, speed, noise level, differential expansion, and others. Much of this dynamic data is taken on line with separate transportable instruments.

Modern computer technology permits a steady state dynamic data collection communication system. By interfacing monitor systems with the computer, raw data can be converted into meaningful information regarding the present condition of equipment and a schedule for the desired preventive maintenance. The development of the hardware for such monitoring systems has been considerable in recent years. The majority of the monitoring has been based upon vibration, temperature, electrical, sound, hydraulic, pneumatic, corrosion, wear, shock-pulse, and dimension and motion patterns.

Monitoring systems not only increase machining uptime, but can also boost productivity during that time. For example, a machine tool equipped with a sensing system that monitors tool condition provides information on the tool's performance, thus allowing the operator to produce parts at a faster rate. Since the system senses changes in performance, which can indicate wear, the operator is able to compensate by changing machine feeds and speeds. The process continues until optimum usage is achieved and the facility is then stopped for tool change. The monitoring system prevents unnecessary tool change and encourages operators to maximize tool usage.

VIBRATION ANALYSIS

The use of vibration analysis in order to learn about an individual machine's condition is becoming increasingly important in connection with those inspections introduced in a PM program. Baseline or signature vibration readings must be made on specific machines when new or newly repaired (overhauled). One facility's signature is not transferable to another machine or other family of machines. The maintenance analyst must identify a periodic check interval for those facilities selected to be included in the program. This interval depends upon several factors. First, the type of machine has a significant impact on the desired frequency. High-speed equipment invariably requires more frequent checks than low-speed facilities. Another consideration depends on the use of the equipment; correspondingly, the interval can vary from daily to monthly.

Periodic checks of overall vibration and velocity levels at selected bearings located in the equipment (including the motors) establish a point where action is to take place. Action involves a vibration analysis in order to identify the problem that has caused the vibration to exceed the limit. As a component of a facility degrades, the peak value at its signature frequency increases, thus providing hard data for estimating when a failure will occur (see Fig. 9.1).

The vibration analysis involves taking overall vibrations horizontally, vertically, and axially at each bearing housing using a vibration meter. In addition, a vibration amplitude-versus-frequency analysis using a vibration analyzer

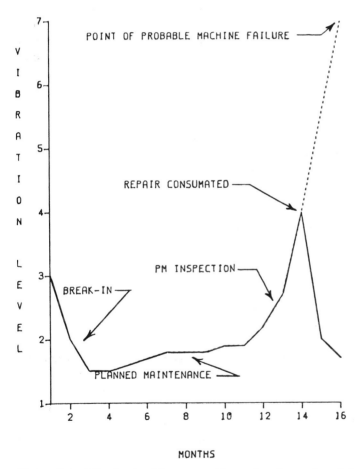

Figure 9.1 Vibration level increases with months of operation of the equipment. The time to perform the necessary repair to avoid breakdown can be predicted.

should be made. Finally, an analysis of the facility condition should be made after the vibration data is evaluated. These data provide a baseline against which future measurements can be compared.

The number of checkpoints depends on how critical the facility is to the plant operation. Once each checkpoint has been determined, it is a good idea to mark it with a dab of colored paint so that future measurements will be made from the exact same point.

Vibration analysis can determine both mechanical defects and failed components, and isolates causes of equipment malfunction.

Instruments for measuring frequency, amplitude, velocity, and acceleration or wave slope are well developed. The accelerometer is an electromechanical transducer which produces an electric output proportional to the vibratory acceleration to which it is exposed. Thus it is a high natural-frequency instrument suitable for general purpose vibration pick-up. Both velocity and displacement for harmonic motion may be obtained from accelerometers when accompanied by an electronic integrator.

On-line vibration monitoring hardware is available to alert when a significant change in vibration has occurred. Upon learning that a change has taken place, the vibration analyzer can be used to perform a series of tests that will identify the specific source of the vibration.

Setting Up the Vibration Measurement Program

The following points should be adhered to in order to have a viable vibration measurement effort:

1. Select the machines to be included in the program.
2. Select and mark check points.
3. Select the periodic check interval.
4. Establish vibration limits. For new equipment, use manufacturer's recommendations. These should be corrected with experience.
5. Conduct a vibration analysis of each selected machine.
   a. Take overall vibrations horizontally, vertically, and axially.
   b. Make vibration amplitude-versus-frequency analysis.
   c. Make an analysis of the condition of the facility after the data is evaluated.
6. Begin a data bank and maintain records of pertinent information relative to all capital equipment included in the vibration monitoring effort.

PROCESS PARAMETER MONITORING

Several of the parameters relating to the efficient operation of a facility or system can be monitored. These include electrical measurements, sound intensity (including ultrasonic) measurements, and hydraulic and pneumatic testing.

The importance of process monitoring can be illustrated by a supply pump used in the oil industry. Obviously such a pump would be considered critical equipment and it is important to know the efficiency of the pump. Regular vibration-based predictive maintenance will provide the mechanical condition of the pump and infrared imaging will provide the condition of both the bearings and the electric motor, but neither of these monitoring procedures will give specific information as to the efficiency of the pump.

In order to determine the pump efficiency, the process parameters of suction and discharge pressures and amp load of the pump need to be monitored. Then the efficiency can be calculated easily.

Electrical Measurements

A significant change in power demand and resistance of electrical components are indications of potential problems. Monitoring of these parameters is a proven technique in connection with predictive maintenance. A diagnostic technician will use several electrical measurements to measure electrical characteristics such as DC voltage, AC voltage, resistance, DC current, and other physical quantities which have been converted into electrical variables such as decibel and temperature measurements. For example, a restriction in a heating duct may be identified by a change in amperes in the fan motor.

Most of these measurements can be made with compact sensitive multimeters. These instruments are small, light, reliable, and relatively inexpensive. They are one of the most important instruments utilized by the diagnostic technician. Motors, for example, need to be inspected periodically to check the installation resistance on the stator and rotor. This instrument can also be used to detect grounding. This inspection is done with a megohmmeter or multimeter. Continuity checks with a multimeter will need to be made on electrical conductors. These can be bare wires, angles, enclosed insulator bars (that accept sliding shoe or roller collectors), or enclosed festooned solid wire. Two other useful electrical tools are as follows:

1. *Logic Monitors and Probes.* Integrated circuits may be monitored by oscilloscopes; however, in view of their size, cost, and required time to set up, they have limited use for the maintenance diagnostician.

Handheld logic probes and benchtop multipoint logic monitors are available that can tell the technician exactly what is happening at any one point in a logic system. By using a digital pulser to initiate a methodical stimulus, a response evaluation of specific sections of circuitry can be made. Thus entire logic trees can be followed and predictive results can be compared with observations. In this way faults may be isolated rapidly and repairs made.

2. *Programmable controllers.* A programmable controller is a digital device that can store a problem in its memory. It can be a valuable tool to the diagnostic and maintenance technician. The programmable controller has three main components: a power supply, a central processing unit, and input/output modules. Thus it can be seen that the programmable controller is a form of a computer. In operation, the central processing unit (CPU) scans the input/output modules, and records the status of each one in its memory. Then, based on each scan, the CPU may initiate one or several actions depending on the information acquired on the I/Os.

For example, a programmable controller can be used to control and monitor almost any process. As a facility begins to wear or produce parts where tolerances are larger than normal, the equipment will automatically shut down or be identified as needing maintenance attention by turning on a service light.

The programmable controller can also be used for energy conservation by turning on and off parking lights that are indexed with time variations for sunrise and sunset. Or it can be used to compare energy usage to an ideal rate and shed an overload during a demand interval in order to avoid a higher demand charge. Later the load will be restored when the demand will not exceed the ideal limit.

Sound Intensity

In many cases, a significant variation in sound intensity from the norm will provide a convenient and reliable indicator of the need for component replacement or equipment overhaul. Sound is a vector quantity that describes both the amount and the direction of net flow of acoustic energy at a given position. The dimension of sound intensity is energy per unit time per unit area. Sound intensity measurements identify not only the amount, but also the exact location of the sound. The maintenance diagnostic technician can identify sources of impending breakdown or malfunction by searching for excessive noise in connection with rotating or reciprocating parts of machinery and equipment. By moving a handheld probe from a real-time sound intensity analyzing system in the area of interest, points of high intensity are identifiable.

The inspector in many cases will be able to identify impending trouble by locating the general source of noise by merely listening. The precise source of the noise may be determined by moving a metallic object such as a screwdriver along the equipment, thus transmitting the uncharacteristic sound to the inspector's ear.

Vibration pickups or electromechanical transducers described earlier can be used to locate sources of noise. These can be used in place of the sound-level meter or microphone such as in the stethoscope and hydrophone.

The reader should understand that every type of moving part sounds different to some degree. Thus, in a complex system such as a robotic controlled machine tool or an automobile or aircraft, almost any function can be distinguished by its unique sound. Each constituent in any kind of rotating or reciprocating part will have its own acoustic signature, referred to as a voice print, which is analagous to a fingerprint.

Acoustic monitoring systems can analyze voice prints of operating systems and separate good products from bad ones prior to shipment thus assuring customer satisfaction and providing at an early time the desirability of initiating maintenance before damage is done.

Automated noise monitoring systems are able not only to identify a trouble site and the faulty part, but can also describe the problem. Thus operators can be alerted to potential failures before they occur, allowing time for planned and scheduled maintenance.

Unlike vibration monitoring, ultrasonics monitor the higher frequencies produced by the unique dynamics of facilities and process systems. Acoustic waves greater than 20,000 cps are referred to as ultrasonic. Low-amplitude waves find considerable application for nondestructive testing. Typically, the resonance system is used. Here, a single transducer, with variations in the frequency applied to it, comprises the operation of the system. The resonance system can determine thickness measurements when only one side is available. For example, it is possible to measure wall thickness and consequently the amount of corrosion of vessels. Since these thickness measurements can be made accurately from one side of the tank wall, they are completely nondestructive.

Low-power ultrasonic waves also have application in diagnostics through the location of cracks and flaws in operating equipment. This ultrasonic system is referred to as *through-transmision*. Here, continuous-pulsed or modulated waves are applied to a transducer couple to one side of the part being inspected with pickup on the other side. If a crack or flaw interrupts, the waves do not penetrate through the part. This nondestructive method of inspection has considerable application for the preventive maintenance inspector. The principal application is in connection with leak detection in valves, steam traps, piping, and so on.

Hydraulic and Pneumatic Testing

Periodic testing of the hydraulic power system in mechanical equipment is always cost-effective. Power monitoring of both hydraulic and pneumatic equipment can identify if there is a leaky valve, if a pump is slipping, or if some other leakage is taking place resulting in a slowdown or erratic operation of the equipment.

In addition to monitoring the power in hydraulic systems, the condition of the hydraulic oil should be regularly checked. Also the condition of the filters should be inspected regularly.

A pressure flow meter is to the hydraulic system what the multimeter is to the electrical system. When the pump flow at rated load becomes unacceptably inefficient, scheduled maintenance should be planned. Pneumatic and hydraulic systems operate within a range of pressures and flows. By monitoring these parameters and analyzing the significant changes, predictive maintenance can be efficiently scheduled.

Fluid level detectors for hydraulic reservoirs can facilitate the regular inspection of hydraulic power systems to assure there is an adequate supply of hydraulic fluid. The moisture level in pneumatic supply systems should also be monitored.

Temperature Monitoring Including Thermography

Elevation in temperature frequently is indicative of impending trouble. For example, motors fail in many ways, but it usually is excessive heat that does the damage. This is often caused by antifriction bearings because the service life is limited. Bearing life is strongly influenced by its mountings and operating conditions and by the preventive maintenance it receives. In order to avoid bearing breakdowns and to plan for their replacements, the equipment using the bearings can be monitored by on-line monitoring of temperature. Typical parameters used to check the condition of bearings while in service include temperature, vibration, noise, and mechanical impacts.

In addition to monitoring bearing areas by temperature variation, the preventive maintenance inspector will want to verify that certain process areas that depend upon given temperature levels, such as furnaces and heating pots, are actually maintaining the temperature desired.

In the case of motors used on critical equipment, it is usually wise to provide each motor with built-in temperature detectors. These can either turn on an alarm or shut off the motor. The following instruments can prove helpful to the inspector for monitoring and controlling temperatures.

1. *Mercury thermometer.* Since most liquid and solid-state materials, including mercury, expand with increasing temperature, a change in temperature will result in a change in length. Accurate calibration can provide a variety of thermometer designs that can give accurate measurement of the temperature as long as the thermometer can be placed in the immediate confines of the heat source. The most commonly used temperature measuring instrument is the mercury-in-glass thermometer where the range of temperature measurement is approximately from -35 to $900°F$.

2. *Bimetallic element.* Bimetallic temperature gauges (used in thermostats, for example) are made by welding together two strips of metal (often invar and brass) having different coefficients of expansion. With a change in temperature the element will distort an amount that is proportional to the temperature.

3. *Gas thermometer.* Since gases expand with increases in temperature, the pressure p of an (ideal) gas will be directly proportional to its absolute temperature T. Increasing temperature will cause a fluid housed in a bulb to increase in pressure thus applying force to a spring to which is attached a pointer on a scale.

4. *Vapor-bulb thermometer.* This thermometer is based upon the rate of change of vapor pressure with temperature. The vapor pressure of all liquids increases with temperature.

5. *Thermocouple.* Thermocouples consist of two pieces of dissimilar metal (usually wire) that are joined together at one end. A voltage is developed whose magnitude is dependent upon the metal used and the temperature of the junction. The small voltage generated at this hot junction will be nearly proportional to the temperature difference between the other end (cold junction) and the hot junction. Thermocouples provide accurate measurements (within 0.5%) up to about $1400°F$.

6. *Thermistor (resistance thermometer).* Thermistors are based upon a variation of electrical resistance. With a change in temperature, electrical conductors will encounter a change in resistance. A Wheatstone bridge circuit is used to make the measurements. The resistance bulb generally utilizes either a platinum or copper wire coil sealed in a protective metal tube.

7. *Radiation pyrometer.* This temperature measurement technique is particularly useful for the measurement of very high temperatures (above 1000°F) where the intensity within a narrow wavelength band can be measured. It is based upon a change in radiation and the fact that a material will radiate energy in proportion to the fourth power of its absolute temperature. Thus the radiation pyrometer is a radiation sensing element consisting of several thermocouples connected in series and utilizing a lens for focusing radiant energy onto a thermo-sensing element. The temperature rise of the element depends upon the total radiation received, and the radiation relates to the temperature of the target being measured. The sensing element often is a thermopile consisting of several thermocouples connected in series arranged so that all the hot junctions lie in the field of the incoming radiation.

8. *Optical pyrometer.* The optical pyrometer is useful for the measurement of very high temperatures (1000 to 5000°F). Here the intensity of the radiation is compared optically with a heated filament. The filament brightness can be varied by a temperature adjustment or a constant brightness filament (by maintaining a constant current flow) and can be compared with the source, which is viewed through a calibrated lens (optical wedge).

9. *Tempilstick (Seger cone).* Here the temperature-sensing element is a solid which will change either color or shape at some critical temperature. Today commercial products are available for measuring temperatures from as low as 120°F to approximately 3500°F.

10. *Electrical systems.* Systems based on the thermocouple and resistance thermometer are applicable where several different temperatures are to be measured. The resistance thermometer is used for low temperatures, while the thermocouple has application in connection with high temperatures. The thermocouple voltage is measured by either a potentiometer or a deflection type millivoltmeter.

    The potentiometer balances the unknown voltage (thermocouple voltage) developed by the primary measuring element against the known voltage from a battery within the instrument that is precisely calibrated using a slide wire. A change in temperature at the measuring element causes a condition of unbalance in the electrical bridge circuit. The amount and direction of unbalance are measured and the slide wire is repositioned to effect balance. Reposition of slide wire or contact moves the pointer or pen to indicate or record the temperature.

11. *Microprocessor-based infrared system (thermography).* The microprocessor-based infrared temperature measurement system makes

possible many applications that would be both difficult and impractical with contact instruments. Infrared emissions are the shortest wavelengths of all radiant energy and are invisible without special instrumentation. The intensity of infrared radiation from an object is a function of its surface temperature. When a small sensing head is aimed at the object being measured, the microprocessor calculates the surface temperature for display and provides output for recorders and controllers. The instrument can replace commonly used temperature sensors in process monitoring and control. For example, infrared monitoring of electrical switch gear has proven successful in predictive maintenance of equipment employing switch gear. Infrared photography has been used successfully to predict refractory life of glass furnace walls, detecting heat loss in pipes, heat loss through walls, roofs, insulated tank cars, and so on.

It can be appreciated that this inspection equipment can identify increases in temperature in hardware with no service interruption. Housings or covers need not be removed in order to measure accurately the magnitude of temperature rise, thus pinpointing problem areas.

This equipment has special application in connection with preventive maintenance inspections of special equipment whose downtime could cause high production losses, spoilage, or hazardous conditions.

Tribology

Diagnostics based on tribology provide information related to the bearing-lubrication-rotor support structure of the equipment being monitored. Two tribological techniques find application in connection with predictive maintenance: lubricating oil analysis and wear particle analysis.

Lubricating oil analysis, as brought out in Chapter 8, involves the taking of samples at scheduled intervals in order to determine the condition of the lubricating film that is important to the efficient operation of the machine. The oil is tested for viscosity, fuel dilution, solids content, oxidation, contamination, fuel soot, total acid number, and total basic number.

Wear particle analysis, often referred to as ferrography, can provide important information regarding the amount of wear that has occurred in the machine train. A relatively new well-maintained machine will contain low levels of solids—smaller than 10 μm. As the machine's condition degrades, the number and size of particulate matter will increase. Wear particle analysis will provide information as to the quantity, size, composition, and shape of solids developed through wear. For example, particles picked up by lubricating oils from a gear box may contain a mixture of particles resulting from normal rubbing wear and cutting wear. The cutting wear particles may be curved, suggesting some alignment problem.

Visual Inspection with Dimension and Motion Pattern Monitoring

Visual inspection implies the utilization of videotape facilities, stroboscopic motion analyzers, lasers, and optical alignment equipment. With these procedures, dimension and motion pattern monitoring can take place on all critical equipment. For example, wear may be identified and measured easily in such moving parts as gears, rolls, shafts, and so on. Similarly, corrosion in pipes, vessels, and reactors should be monitored so that planned maintenance can take place at an optimum time.

1. *Videotape system.* Videotape equipment, developed in recent years, offers the PM inspector a useful and powerful tool to facilitate the diagnosis of operating equipment. Videotape systems can record up to 8 hr of facility operation. The tapes can be run in slow motion without loss of picture quality. Speeds can be varied from ¼ to ⅟₃₀ of normal, allowing thorough analysis of moving equipment for misalignment, excessive travel, eccentricity, and similar characteristics that can lead to failure or shutdown. The camera and recording equipment is able to freeze one field at a time to provide sharp, jitter-free still pictures.

    Most modern video camera systems include an integral stopwatch function with the camera control. This can be displayed in any corner of the screen, providing elapsed time in hundredths of a second.
2. *Stroboscopic motion analyzer.* Often it is useful to view machine parts that are rotating or reciprocating at high speeds in a slow motion image. In this way the diagnostic technician is able to follow the movement of a facility component through its complete cycle and evaluate its motion pattern with what is considered normal. The stroboscopic motion analyzer permits this type of examination.
3. *Laser.* Laser beams can be used to measure accurately straightness, parallelism, squareness, and flatness. Velocities up to 720 in. per minute can be followed. Laser systems are being used by the maintenance technician to calibrate accurately lead screws, to align the ways of machine tools, and to measure misalignment and variation of straightness.

4. *Optical alignment.* Light can be used as a standard for the measurement of such properties as distance, straightness, and squareness. Today there are several optical instruments available to the diagnostic technician for obtaining precise measurements. For example, the measuring microscope is used for measuring very small dimensions such as small displacements. The optical comparator allows the technician to project a magnified shadowed image of an object on a screen where it can be compared with a reference template.

Other optical equipment that can be used to establish and test alignment, squareness, and flatness include alignment telescopes, collimators, jig transits, and optical squares.

Shock-Pulse Measurement

The shock-pulse measurement technique reported by Testing Machines Inc. of Amityville, New York, and SPN Instrument Co. of Sweden has application in determining the operating condition of antifriction bearings. This technique measures the magnitude of a mechanical impact (the ball of an antifriction bearing when it encounters a flaw in the bearing race and drops to the bottom) by measuring the shock pulse. By measuring these impacts, the inspector can follow the change in a bearing's condition and can estimate when it reaches that point when it should be replaced. Thus the inspection allows the prediction of when maintenance should take place and this date can then be scheduled.

The instrumentation utilized includes a transducer where a dampened resonant oscillation is established when the shock hits it. The amplitude increase of the oscillation is determined by an increase in the pressure wave front resulting from the mechanical impact of the balls within the bearing race.

NONDESTRUCTIVE TESTING

There will be numerous occasions, especially in connection with mechanical equipment, where the PM inspector will need to check a component for flaws, cracks, and so on, due to fatigue, misuse, or wear. Usually it is essential that the component be inspected without destroying it. Several nondestructive testing and inspection techniques are available. The most important of these are magnetic-particle, radiographic, sonic and ultrasonic, and fluorescent penetrant oil inspection. Sonic and ultrasonic procedures have already been discussed. The other nondestructive methods will now be summarized briefly.

1. *Magnetic-particle inspection.* At times, the diagnostic technician will want to inspect components for structural discontinuities devel-

oped through service. For ferrous components (except austenitic steels), magnetic-powder tests—commonly known as magnaflux—are useful. In this test, a magnetic flux is induced in the part being inspected. Then the part is surface dusted with finely divided particles of ferromagnetic materials that offer a low reluctance path to the leakage field and take a position that outlines the location of any surface cracks. The excess powder is blown off, and if a defect exists it is readily discernible by the remaining powder particles.

The magnetic powder may be applied either when the magnetizing current is flowing or after the magnetic flux is induced. The magnetic powder can be applied either dry or wet (suspended in a petroleum distillate).

2. *Radiographic inspection.* Both x-ray and gamma ray examination may be used by the PM inspector in connection with nondestructive testing and inspection. The use of x-rays has been limited to approximately 12 in. of penetration of steel at 2000 kV while gamma rays have been used for inspecting steel more than 16 in. thick. However, x-ray inspection is much more rapid than gamma ray inspection and is superior in finding defects in sections up to 2 in. thick. If there is considerable variation in the thickness of the component being inspected, gamma rays may be preferable to x-rays since they scatter less.

When using radiography for the locating of defects both photographic and fluoroscopic techniques may be used. In both x-ray and gamma ray photography, film is placed close to the component being inspected so as to intercept the radiation that is passed through the selected area of the part. The resulting photographs made by means of x-ray radiation are referred to as *exographs* and those produced by gamma radiation are called *gammagraphs*. Fluoroscopic methods are especially applicable where no permanent record is needed and rapid inspection is desired.

3. *Fluorescent penetrant.* Under this method of nondestructive inspection, a fluorescent penetrant oil is applied to the test part. The oil is then washed from the part surface by a stream of water. The fluorescent film that penetrates the surface defects will not be washed away since water has poor penetration properties. The part being inspected is now dried, and a developing powder is distributed over its surface. The powder draws the penetrant oil from the fine cracks, dispersing it on the part surface. Fluorescent indications of any defects appear when the part is inspected under a light that emits near-ultraviolet light. This light is commonly referred to as black light.

INSTRUMENT AVAILABILITY

Several of the instruments described in this chapter are costly and may be difficult to acquire by the small or medium-sized maintenance department. This should not discourage their use in connection with diagnostic procedures used in conjunction with preventive maintenance. Frequently, these instruments are already being used in the inspection department. If so, it should be a simple matter to have them periodically loaned to maintenance. Often it is wise not only to borrow the instrumentation from inspection, but also the services of the technician who is skilled in the operation and calibration of equipment. Such a cooperative plan allows the maintenance operation to borrow both instruments and knowledge from other departments, such as inspection, research, and production, when these departments happen to have the desired instruments and the skills and experience to use them.

When the necessary instruments are not available internally and their services are only required occasionally, it may be advantageous to rent them from a neighborhood industry or business. Again, it may be desirable to rent both the equipment and the technician who operates the equipment. For example, let us assume that a critical machine tool is suspected of having an interior defect in one of its components. It would be desirable to x-ray this forged part to determine its present quality. The company has no portable x-ray unit to inspect the part and certainly does not want to purchase a unit for this isolated inspection. Perhaps a neighboring industry that does a great deal of nondestructive testing would be glad to loan or rent one of their portable x-ray units for a period of an hour to perform the inspection.

TRAINING FOR DIAGNOSTICS

The diagnostic technician is a highly skilled individual who has had extensive training and experience. The availability of this type of technician, especially in smaller plants, is limited. As mentioned, in small and even in middle-sized plants, it will probably be cost-effective to procure the services of a diagnostic technician from some neighboring operation that maintains this class of personnel.

Those technicians who perform diagnostics will find that it is necessary to continually update themselves in order to stay abreast of the ever-increasing new technologies characteristic of modern facilities and equipment. In larger plants, it is wise for the company to offer in-plant training courses regularly to update those maintenance personnel associated with diagnostics and to train the newer employees. Much of the instruction can be provided by engineers from the equipment manufacturers and from instrument producers. Often engineering talent within the company can be utilized to handle a portion of the instruction.

It is important that the training includes not only theory but hands-on experience with the hardware and instrumentation that is utilized. Details of such in-plant training programs are discussed in Chapter 15.

SUMMARY

The ability to diagnose the condition of key equipment is the heart of predictive maintenance. As production equipment becomes more sophisticated, so the instrumentation needed to diagnose it becomes more complex. It is essential that technicians associated with predictive maintenance receive regular training in the use of modern instrumentation for sensing those parameters that facilitate the diagnosis of critical production facilities used in a plant or business (see Chapter 13).

In connection with electronic equipment, including robots, where many sensors are incorporated in the design of the equipment, it usually is cost-effective for the small and medium-sized operation to use the service function of the equipment manufacturer rather than train their own personnel in the intricacies associated with the equipment. Thus standby equipment, in connection with critical facilities, may be an economic alternative as opposed to regular costly diagnostic procedures utilizing sophisticated instrumentation and highly skilled personnel. When such equipment does fail, the standby equipment goes into line while the equipment manufacturer services the downed facility.

However, the vast majority of mechanical, electrical, hydraulic, and pneumatic equipment that may be classed as critical in the operation of a plant should be subjected regularly to those diagnostic techniques that can best identify their current condition and estimate when repair, modification, or overhaul should take place to avoid crashdown. These inspections and evaluations are usually performed best by internal technicians that have been trained properly.

SELECTED BIBLIOGRAPHY

Corry, Robert T. *Instruments and Controls, Section 16, Marks' Standard Handbook for Mechanical Engineers*, 8th ed. New York: McGraw-Hill, 1978.

Bruel and Kjaer. *Instrumentation for Sound, Vibration, Illumination, Thermal Comfort, Medical Diagnostics and Signal Analysis*. Marlborough, MA, 1983.

Nakajima, Seiichi, Yamashina, Hatime, Kumagai, Chitoku, and Toyota, Toshio. "Maintenance Management and Control." In Gavriel Salvendy, Ed., *Handbook of Industrial Engineering*, 2nd ed. New York: Wiley, 1992.

10
Computerized Maintenance

Computer systems are an integral part of modern engineering maintenance management. In order for the whole maintenance system to function effectively, five principal computer systems should be designed correctly and operated. These five systems are

1. Equipment control
2. Work control
3. Maintenance spare parts and inventory control
4. Cost accumulation and reporting
5. Performance reporting

These five computer systems control the three principal variables (equipment and facilities, labor, and materials); accumulate and summarize costs; measure performance in comparison to some standard; and report to management the effectiveness of the entire integrated management system.

EQUIPMENT AND FACILITIES CONTROL

The equipment register is a computer data base created from a coding sheet (see Fig. 10.1). A typical printout that illustrates the completeness of the equipment listing is illustrated in Fig. 10.2. The equipment register that is

EQUIPMENT CODING SHEET

CATALOGER M CARNAHAN   DATE [ ][ ][ ]   PAGE [1] OF [1]

EQUIPMENT NUMBER 0304 1 KGT301   ASSEMBLY NUMBER

F S A	COMMODITY NO. (10)	MANUFACTURER'S SERIAL NUMBER (20)	ASSET NUMBER (10)
1 A A	0234 11 2243 82		M788 86465 1

	NOUN NAME (20)	SHORT DESCRIPTION (35)
2 B A	TURBINE, GAS	INJECTION COMPRESSOR GAS TURBINE

LONG DESCRIPTION (55)

3 C A	TURBINE, GAS
3 D A	HP (AT 1.0 FEET ALT): 27950
3 E A	TEMP (IN): 90 DEG-F TEMP (OUT): 956 DEG-F
3 F A	PRESSURE (IN): 14.56 PSIA PRESSURE (OUT): 14.81 PSIA
3 G A	CYCLE: SIMPLE SHAFTS: 2
3 H A	ROTATION: CCW CASING SPLIT: HORIZONTAL
3 I A	RPM (TURBINE): 4670 RPM (COMPRESSOR): 5100
3 J A	STAGES (TURBINE): 2 STAGES (COMPRESSOR): 16
3 K A	CONTROL SYSTEM: SPEEDTRONIC
3 L A	STARTING MOTOR: GE CUSTOM 8000 ELECTRIC
3 M A	GE MODEL SERIES: MS-5002-B
3 N A	
3 O A	DRIVES: INJECTION COMPRESSOR - DRESSER/CLARK - 272B4/4
3 P A	DRIVEN EQUIPMENT NO: K-301
3 Q A	

NARRATIVE LOCATION (55)

| 3 R A | CLUSTER 111 - ADJACENT TO INJECTION COMPRESSOR |
| 3 S A | |

	MANUFACTURER'S NAME & ADDRESS (30)	VENDOR'S NAME & ADDRESS (25)
4 T A	GENERAL ELECTRIC CO.	
4 U A	GAS TURBINE PRODUCTS DIV.	
4 V A	SCHENECTADY, NY 12345	
4 W A		
4 X A		

	MFG. NO.(5)	VND. NO.(5)	P.O. NUMBER (10)	P.O. DATE (6)	ITEM	UNIT COST (8)	CRITICALITY
5 Y A		62 10KGT3	091776		00000 001

Figure 10.1 Typical coding sheet.

EQUIPMENT NO.   5030411 KGT 301   TURBINE, GAS   INJECTION COMPRESSOR GAS TURBINE - 27950 HP

IDENTIFIERS   MANUFACTURER   LONG DESCRIPTION
  Assembly/Inst.   Gas Turbine Product Div.   Turbine, Gas
  Commodity Code   37 30 11 000   General Electric Co.   HP (at 10 ft. alt.): 27950
  Mfg. Serial No.   244382   Schenectady, NY 12345   Temp. (in): 90 deg.- F. Temp. (out): 956 deg. - F.
  Co. Asset No.   M788864651   Pressure (in): 14.56 psia Pressure (out): 14.81 psia
  Bech. No.   KGT - 301   Cycle: simple   Shafts: 2
                                                                      Rotation: C C W   Casing split: Horizontal

P.O. INFO.   VENDOR
  Purchase Order   62.10 - KGT - 3   Gas Turbine Products Div. RPM (Turbine): 4670   RPM (Compressor): 5100
  Date   09/17/84   General Electric Co.   Stages (Turbine): 2 Stages (Compressor): 16
  Line No.   1   Schenectady, NY 12345   Control System: Speedtronic
                                                                      Starting Motor: GE Custom 8000 Electric
  Criticality   1   GE Model/Series: MS - 5002 - B
                                                                      Drives: Injection Compressor - Dresser/Clark

LOCATION   CLUSTER III   ADJACENT TO INJECTION COMPRESSOR

Figure 10.2 Computer printout of equipment listings.

maintained consists of four parts: a master record, lubrication specifications, a repair summary, and a time and cost estimator. The master record as shown in Fig. 10.1 is maintained for all equipment.

The lubrication specification provides for every lubrication point, the name of the lube, the method of application, the change frequency, and the change volume. Then a record of those items of equipment and facilities that require lubrication service each day can be entered in the computer in the form of string instructions for each facility. A printout of these lubrication instructions is made in ample time to schedule efficiently this preventive maintenance work. The printout will provide information as to facility location and type of lubricant (lubricants) to be used.

The following program illustrates the simplicity of providing such a daily lubrication schedule.

```
5    CLEAR 100: CLS
10   REM PROGRAM FOR LUBRICATION
20   PRINT "ENTER THE FIRST TWO LETTERS OF THE DAY"
25   PRINT "ENTER TODAY'S DATE, SEPARATE FROM THE DAY BY
     COMMA"
30   INPUT D$, Y$
40   FOR I = 1 TO 5
50   READ D1$, M1$, M2$, M3$
60   IF D1$ = D$ THEN 130 ELSE 70
70   NEXT I
80   DATA MON, VER MILL, RADIAL DRILL, TURRET LATHE
90   DATA TUE, HOR MILL, THD GRINDER, CENTER GRINDER
100  DATA WED, INT GRINDER, ENGINE LATHE, BROACH
110  DATA THUR, CIN MCH CENTER, N/C AMADA, PUNCH PRESS
120  DATA FRI, N/C MILL, HONING, N/C LATHE
130  CLS: PRINT D1$, Y$
140  PRINT "LUBRICATION FOR THIS DAY IS ON"
150  PRINT M1$
160  PRINT M2$
170  PRINT M3$
180  INPUT "TYPE YES FOR ANOTHER PRINTOUT"; C$
190  IF C$ = "YES" THEN 5
200  END
```

(Adapted from program in Nicks, J. E., *Basic Programming Solutions for Manufacturing*, Dearborn, MI: Soc. of Mech. Eng., 1982.)

The repair summary provides a list of standard jobs and the best procedure to be followed to do the work along with an estimate of the standard hours required to perform each operation. The time and cost estimator provides an estimated summary of cost based on operations performed and their frequency and cost centers.

An important subset of equipment control that involves computer application is energy conservation. The cost impact of energy is making this subject of considerable importance to competitive business and industry. Energy conservation in most companies is being delegated to plant engineering and maintenance (see Chapter 12).

The computer can be especially helpful in developing inspection schedules for the purpose of energy conservation. For example, steam traps, numbering in thousands in many plants, is a fertile area to study in order to conserve energy. A leaking steam trap can cost the enterprise more than $1500 per year in lost energy (based on steam at $5.00/1000 lb and when steam gauge pressure is 125 psi and orifice diameter is 3/32 in. This results in 36.2 lb/hr steam loss. Then 365 days × 24 hr/day × 36.2 lb × $5.00/1000 = $1586). The computer is an effective tool to control load-cycling, various start-stop mechanisms, lighting systems, and so on, in order to conserve energy. This subject is discussed in more detail in Chapter 12.

FLEXIBLE MANUFACTURING SYSTEMS

A flexible manufacturing system has been defined as "a collection of two or more manufacturing units interconnected with material handling machinery under the supervision of one or more executive computers."* The executive computer usually is programmed to handle the control of the machine operation, the scheduling of parts through the manufacturing system, the accumulation of important data including the estimated tool wear associated with each part program, and failure diagnosis.

Thus the executive computer, under the direction of the trained technician, is capable of interrogating various components and parts of the manufacturing system to assist in location of symptoms of impending failure. When a system fault is detected by the executive computer, it will initiate a command that will stop the operation of the defective manufacturing system component. The executive computer will continue to monitor the entire system on a hierarchy basis during a fault condition. Thus the identification of more than one fault can be accomplished.

*Theodore J. Kuemmel, "Systems Considerations in Flexible Manufacturing," SME Paper MF 81-101.

The system operator can typically interrogate the executive computer through a terminal in order to obtain more details as to the nature of a defect or fault. He can then make a decision as to whether the specific station should be shut down and repaired, bypassed, or left on line while the fault is being removed.

WORK CONTROL

The work control system administers the maintenance work order system. The details of maintenance control systems have been discussed in Chapter 2. At this point, we will see how the flow of the work order information along with the time sheet data is incorporated in the information system controlled by the computer.

We have learned that the work order document should be designed to serve three functions:

1. To authorize and define work to be performed by the maintenance department
2. To plan, perform, and control the actual work
3. To be used as a source document to accumulate maintenance costs

These functions can be achieved only if there is

1. A systematic screening and work authorization procedure
2. Information available so that accurate planning and scheduling can be accomplished, including standard time information
3. Reporting of cost and progress information and accumulating of data so that improvement can take place

The maintenance work order should be assigned a code number so as to break down maintenance cost by major categories for comparison purposes. Figures 10.3 through 10.5 illustrate how one company handles the filling out and planning of the work order. Figures 10.6 and 10.7 show the flow of information for routine and emergency type work orders.

It should be understood that pertinent information from the work request or work order is introduced into the memory of the computer for control, follow-up, and reporting purposes. Of course the maintenance work order itself can be the output of a computerized system. The approved maintenance request is the basis of the input information to the computer where details of the work request are stored. Information can be added readily, changed, or deleted as necessary. The work order, as in a manual system, becomes the information center for work scheduling and tracking as well as outlining what

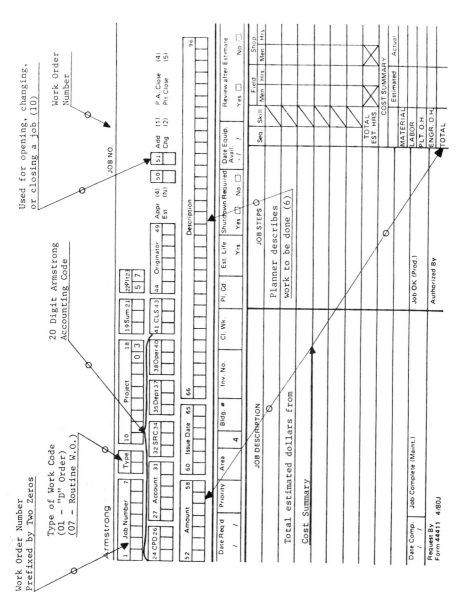

Figure 10.3 Instructions for filling out first and second lines on the work order (responsibility of planning section).

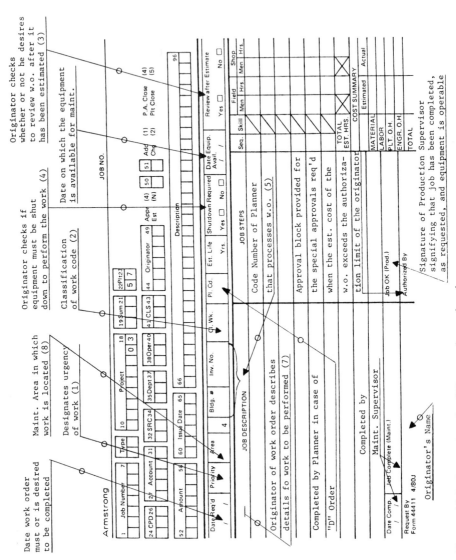

Figure 10.4 Instructions for filling out the work order.

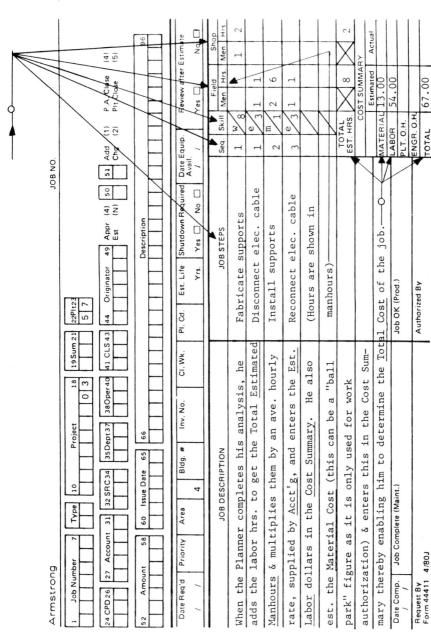

Figure 10.5 Instructions to the planner on how to record his analysis of a job on the work order.

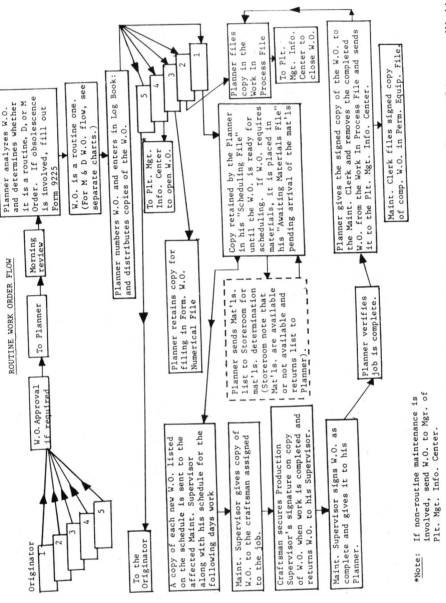

Figure 10.6 Routine maintenance work order flow characteristic of a large company. (Courtesy of Armstrong World Industries, Inc.)

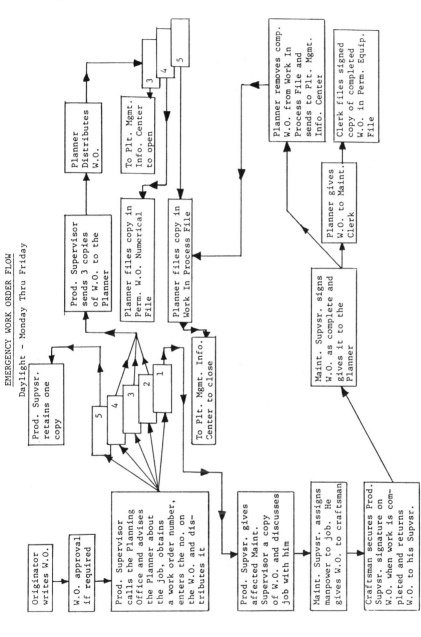

Figure 10.7 Emergency maintenance work order flow characteristic of a large company. (Courtesy of Armstrong World Industries, Inc.)

work should be done, how it should be accomplished, and how much time and material should be utilized. As work requests are received and approved, the planner will input all pertinent information onto the computer screen just as written work orders would be typed or written. When the work orders are scheduled, assigned and completed, additional information is added. Figure 10.8 illustrates a maintenance work order that is the hard copy output of a computer system. Note the hand postings made when the work was completed by the workman (date maintenance work completed, skill code, maintenance work code, and hours to complete) and the cost center (cost of maintenance performed). This hand-posted information subsequently is entered into the computer data base for cost analysis and management reports.

Typically maintenance requests are grouped together in an active file on a storage diskette. Each approved request is referred to as a *record*. As records are completed, they are copied on to a different diskette which has a file of all completed requests over a certain period of time. Removing a completed request from the active file to a completed file will of course provide storage space for new work requests

One computerized work order system uses the following input information:*

1. *Maintenance request number.* This number is assigned automatically when new requests are entered and becomes the label or name for the maintenance request record.
2. *Location code.* This is a two-digit code used to indicate where the work needs to be done. The names of the various locations represented by the codes are kept in a separate file (diskette) so that it is not necessary to have the entire name typed on each request.
3. *Type work code.* A two-digit code is used to separate requests by broad categories such as preventive, routine, alterations and modification, vandalism, contract and capital improvement. Typed work code descriptions are kept in a separate file.
4. *Trade code.* A two-digit code is used to separate requests by specific trade descriptions (e.g., 01 = carpentry, 13 = glass replacement, etc.).
5. *Date requested.* This is a six-digit code that identifies the date on which the maintenance request was received.
6. *Work requested.* Here two lines are reserved, each line providing 38 characters.

*Maintenance Work Order. Center Computer Consultants, State College, Pennsylvania.

MAINTENANCE ITEM WORK ORDER

STOCK NO.: 4520-00234-000056   LOCATION: 12-1-172   MAINT. REQ. CODE: 236   FREQ. CODE: 09

MFG. NAME: Herman Nelson   MFG. NO.: MP 212   MFG. MODEL: 3000310

MFG. SERIAL NO.: 34267   NOMENCLATURE: Univent

TECHNICAL DESCRIPTION: 115 volt 60 cycle 1 ph 1.5 amp.

MAJOR ASSEMBLY: Heating system   SKILL CODE: 1   TITLE: Electrician

MAINT. REQUIRED – DESCRIPTION: Clean filter, calibrate and lubricate as required

MAINTENANCE COMPLETED

DATE MAINT. COMP.: 4/21   SKILL CODE /   MAINT. PERFORMED CODE   124

MAN HOURS: 3   COST OF MAINT: $36.00

REMARKS   Filters will require replacement next PM inspection

Figure 10.8 Work order as the output of a specific computer-controlled system.

7. *Employee number.* A six-digit employee designation. Employee names are kept in a separate file so that the names do not have to be typed on each maintenance request.
8. *Date assigned.* The date when a maintenance request is assigned to a maintenance employee who will do the work.
9. *Date completed.* The date when the work was completed.
10. *Work completed.* Here two lines are reserved with 38 characters on each line to provide a description of the work done.
11. *Total hours.* The amount of time spent to complete the job.
12. *Total cost.* The total cost of a job in terms of labor and material utilized.

The computerized maintenance work order system described here uses five different options to enter information. These are:

1. Add maintenance requests
2. Assign maintenance requests
3. Enter information for completed maintenance requests
4. Update maintenance requests
5. Delete maintenance requests

The first three options are used routinely to enter, assign, and complete maintenance requests. If information is not entered for a field, the field will be left blank until the information is added. The update option is used to change information after a request has been added and assigned. All information in the file can be updated except the maintenance request number. If a maintenance request is to be cancelled, it can be permanently removed from the file by using the delete option.

The maintenance work order main menu is the starting point for all the program options except for data file initialization. The main menu which appears on the screen after entering the appropriate option lists the following ten alternatives:

1. Add maintenance requests
2. Assign maintenance requests
3. Enter information for completed maintenance requests
4. Update maintenance requests
5. Delete maintenance requests
6. Generate maintenance reports
7. List requests on file
8. Purge file
9. Code table information
10. Exit program

This rather elementary computerized maintenance work order program can be the basis of maintenance work scheduling, maintenance planning and budgeting, and accountability reporting.

The program will be able to keep track of work requests and completed work by organizing all pertinent information so that it is readily available. For example, reports of Total Active Maintenance Requests, Unassigned Active Maintenance Requests, and Assigned Active Maintenance Requests can be generated regularly. In the reporting, maintenance requests can be grouped by building or location. Further refinement is readily achieved if desired so that grouping can take place by trade type such as plumbing, electrical, carpentry, and so on. Figure 10.9 illustrates such a weekly report for the Carpentry/Masonry activity of a maintenance group in an elementary school.

MAINTENANCE SPARE PARTS
AND INVENTORY CONTROL

During the work order analysis process, the planner lists the predeterminable materials required for the work to be completed. The stock materials and spare parts that are needed are listed on a requisition form. A separate requisition number is used for each work order involved to accumulate cost information in connection with the computer outputs.

Immediately after the work orders are analyzed, the stock and stores requisitions are sent to the storeroom to confirm that the materials indicated on the computer-controlled perpetual inventory are available. When the required materials are confirmed to be on hand, the storeroom clerk stamps *materials on hand* on the requisitions and returns them to the planner for scheduling of the work.

If in the process of analyzing a work order the planner discovers that some nonstock materials are required, a purchase requisition or make work order is initiated.

Effective spare parts control is critical to the overall performance of the maintenance system. There are several computer algorithms that are available that permit precise control of all maintenance materials and spare parts. Maximum and minimum amounts maintained in inventory as spare parts and service components are planned on the basis of past experience and history of the equipment, correspondence with suppliers, catalogs, and of course economic appraisal based on the cost of inventory and the cost of procurement and resulting delays.

A clear-cut division should be made between repair stock items and spare parts that are carried as insurance against prolonged downtime. Since many

TOTAL ACTIVE MAINTENANCE REQUESTS
08/01/84

REQ NUMBER	REQ DATE	JOB TYPE	WORK REQUESTED	EMPL NUMBER	DATE ASSIGNED
			LEMONT ELEMENTARY SCHOOL		
CARPENTRY, MASONRY					
1	05/01/84	2	Replace the outside door of classroom No. 313A in the south wing.	0154	05/10/84
2.	05/11/84	2	Repair the boy's bathroom stall. It has come loose from the floor. South hall.	0162	06/11/84
5.	07/03/84	6	Build 6 benches for the playground area and install them.	0164	07/08/84
7.	07/28/84	2	Build a platform for the Christmas tree.		
8.	08/01/84	2	Cement the cracks in the pavement at the side school entrance.		

TOTAL REQUESTS: 5

LEMONT ELEMENTARY SCHOOL
TOTAL ACTIVE REQUESTS: 5

Figure 10.9 Computer-generated report of a specific system that summarizes all active maintenance requests.

expensive spare parts are frequently inventoried for several years before being used, and some are never used, the expected costs of equipment downtime while waiting for part procurement and the probability of this occurring must be compared with the known cost of the spare parts inventory. Thus the computer algorithms developed will need to consider not only the time value of money, but also the concept of expected cost developed from the probability of part failure during the life of the equipment.

Repair stock is perpetually controlled based on turnover rates. The average time in inventory of repair stock is usually much shorter than repair stores. Maximum and minimum inventories are maintained based upon economical order quantities for purchase parts and economic lot size for internally produced parts. Maximum inventories consider a safety stock determination. On those items where the greatest control is exercised, the least quantity of safety stock is needed.

As with production materials, an ABC value–volume analysis of all ledger maintenance stock and stores should be made. Ten to fifteen percent of all items will represent the major share of the inventory cost (probably 70 to 85% of the items). These class A items will be carefully controlled utilizing the capability of the computer. Typically, ordering is based on a maximum inventory of a 3-mo supply with a safety stock of 2 wk. The average time spent in inventory will then be approximately 7 wk [i.e., (3 mo × 4 wk + 2 wk)/2].

Class B items represent 20–30% of the total and about 25% of the annual dollar usage. These items are also carefully controlled. Here a maximum inventory of about a 4-mo supply is maintained with a safety stock of about 1 mo.

On class C items a 6-mo supply typically is maintained with a 2-mo safety stock. Perpetual inventory control is not carried out on these items. When the last package, keg, box, etc. is opened (this is usually the safety stock inventory), a requisition is placed to order a 6-mo supply.

Thus under the ABC technique, we are able to manage thousands of items by controlling only several hundred. The computer handles the routine changes in price, lead time, and use rate that could cause a change in order point or order quantity.

Another useful computer output is a critical spare parts listing. This would not only identify the various spare parts and where they are used, but would also show quantities on hand and on order as well as the location number identifying where they are stored. This listing is of particular help in the maintenance department in connection with the making of repairs of an emergency nature.

COST ACCUMULATION AND REPORTING

In order to obtain the details of maintenance cost, there must be proper reporting within the work order and control subsystem. Thus accumulated costs should be charged either to the specific asset on which the work was done or else to the cost center where the facility is located. These costs should not only include the maintenance labor involved but any spare parts, stock, and supplies that are utilized. Expense, or overhead, is also added in order to arrive at true maintenance costs.

It is a simple matter for the computer to accumulate all maintenance costs for a facility or a cost center and provide hard copy of these costs along with similar maintenance costs of other equipment and cost centers of the company. The key to cost accumulation is the machine or cost center asset number. Figure 10.10 illustrates a plant equipment maintenance cost record as printed out by the computer.

PERFORMANCE REPORTING

In Chapter 14, several management reports are discussed. There are many indicators which can be used to evaluate the performance of the maintenance function. It is not necessary for an enterprise to use a large number of indicators, as usually a few will suffice. With computer-controlled subsystems and accurate reporting, the following can be reported regularly as computer output:

1. *Overtime hours/regular hours.* This ratio is important since a high value indicates too many breakdowns and work stoppages that have not been planned, a weak or inefficient preventive and predictive maintenance program, a high percentage of unscheduled maintenance, and high costs. A low value indicates overstaffing and probably high costs. Most managements today feel that this ratio should be between 0.05 and 0.08. The ratio by itself, unless it is very high or very low, is not particularly important. However, the trend is important. The reporting should show what the ratio has been for the past 12 periods (assuming monthly reporting) so that comparisons can be made.

2. *Plant availability report.* This report is especially important to management. Here the various departments throughout the enterprise are compared as to their availability during a period (usually 1 mo). High availability usually is an indication of an effective maintenance program characterized by a sound predictive and preventive maintenance effort (see Fig. 10.11).

3. *Plant monthly maintenance summary.* This report shows the maintenance performed at a variety of cost centers within a given product

JOB NUMBER	DESCRIPTION OF WORK	CLASS OF WK	COMP DATE	LABOR HOURS	LABOR	OVERHD	MATERIAL	TOTAL
EQUIPMENT CLASS 001								
002123107	MODIFY SEAL WTR SUPPLY - DEFLAKER	8	02 - 01					
002172507	REPL OPER ON KRAFT FILL VALVE	8	01 - 11					
002208507	RELOCATE KRAFT FILL VALVE	8		1.0	7	6	51	64
005318001	PROVIDE ALARM KRAFT FILL VALVE	8						
005322801	PROVIDE SPARE IMPELLER - SHAFT	8	12-07					
005339701	INSTALL DRIVE TENS ON KRAFT PULP	8						
064580002	REPLACE KRAFT PULPER IMPELLER	8						
064590002	REPLACE MEASURING TANK	8	01 - 17					
064600002	P&I PIPING - KRAFT MEAS TANK	8	01 - 17					
064619902	REMOVE OLD MEAS TANK	8	01 - 17					
	EQUIPMENT CLASS TOTALS			1	7	6	51	64
EQUIPMENT CLASS 003								
002201907	TEMP RELOCATION PUMP FR KRAFT	8		98.5	678	583	1044	2305
	EQUIPMENT CLASS TOTALS			99	678	583	1044	2305
EQUIPMENT CLASS 006								
064050002	PJ PAPER PULP CLEAN EQUIPT	8		13.0	92	69	4962	5054
064060002	PJ ELECTRICAL EQUPT	8		12.0	80	112	6665	6814
064070002	PJ CLEAN EQUIPT PUMPS, PIPING	8		19	130		503	745
	EQUIPMENT CLASS TOTALS			44	302	181	12130	12613
EQUIPMENT CLASS 501								
001018907	REPACK KRAFT PULPER AGT	1	02 - 01	8.0	56	48		104
001070507	REPR-REPL WIPER BL KRAFT PULPER	1	02 - 15	9.5	67	58	458	583
001073307	REPR TWISTED BELTS AGT DRIVE	1	03 - 08	141	915	765	647	2327
001075407	REPACK-CK BRGS KRAFT AGT	1	03 - 01	10.0	70	61	29	160
001079407	REPR BELT DRIVE KRAFT PLPR AGT	1						
001081607	REPR WIPER BL K PULPER AGITATOR	1	03 - 15	2.0	14	12		26

Figure 10.10 Plant equipment maintenance cost record report as printed by the computer.

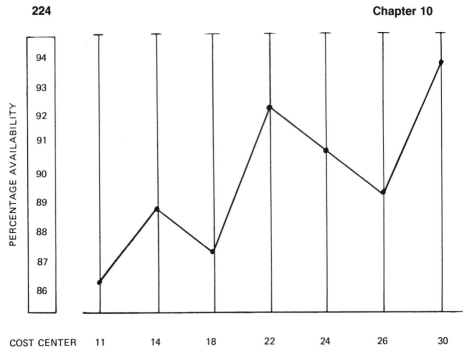

Figure 10.11   Cost center availability report for month of March.

classification such as "Kraft pulping and refining," "board mill," and so on. The amount of maintenance performed at any cost center for the current month can be compared with the amount performed during any month during the past year (see Fig. 10.12).

4. *Estimated-versus-actual hours report.* Figure 10.13 illustrates a computer output showing the total cost of current maintenance jobs and the estimated hours to perform the work compared with the actual hours required.

5. *Distribution of maintenance work by priority level.* Figure 10.14 shows the distribution of maintenance work by five priority levels (E, 1, 2, 3, and 4).

6. *Backlog report.* Figure 10.15 illustrates a work order backlog record for millwrights. Note that this report indicates an excessive backlog indicating the need to add additional manpower of the millwright classification.

Other reports that can readily be developed with the use of the computer include estimating efficiency, completed jobs, trend, and average number of maintenance man-hours utilized.

ME545   PERIOD MARCH   PENSACOLA PLANT MONTHLY MAINTENANCE SUMMARY   PAGE 5

EQUIP. CLASS	DESCRIPTION	VARIANCE MONTH	VARIANCE YTD	TOTAL YTD	MAR	FEB	JAN	DEC	NOV	OCT	SEP	AUG	JUL	JUN	MAY	APR
	OPERATION 070 (CONTINUED)															
	ELECTRIC POWER															
399	MISC BLD'G REPAIRS			0											240	
536	CONTROL CENTERS MTR.	196		1223	196	779	434	297	17	21	47	54	67	247	131	616
599	MISC MACHINERY REPR				9								314			
641	SUB-STATION, ELECTRIC			0						240			89			
	OPERATION TOTAL	509	723	1419	205	779	434	297	17	683	47	143	470	247	376	1341
	OPERATION 080															
	KRAFT PULPING AND REFINING															
000	*			0							30					
001	MACHINERY EXPENSE			569	199	370		239	209	300		1746	136	2081		
003	MACH REL/REM EXPENSE			0						233	34	174		83	2132	
067	*			0												
399	MISC BLD'G REPAIRS			0						182		1119				
401	LIFT TRUCK NO 1			0												
466	TRUCK-DUMP			16			16									
501	AGITATOR-SIDE ENTERG			1513		155	1357		532		244		1491	1186	2101	1868
502	AGITATOR-TOP ENTERG			0							130		476	751	374	34
532	CLEANER CENTRIFUGAL			113			113		1140	391		415	345	1245		
537	CONTROLS-ELECTRICAL			142								78				
544	CRANE-HOIST-ELEVATOR			46	142		46					37				
551	DEFLAKER			5604	531	861	4212	10	46		5096	4241	159		4	2296
554	DRIVE-FIXED OUTPUT			84											435	
575	GUARDS-EQUIP. DRIVE			3534	9	1213	84	143		996	846	1148	165	291	111	216
580	HYDRO-PULPER			42	17	17	2312	13		21	59	17	634		69	4
584	INSTRUMENT & METER			0			8			1067	1352	68		475	1286	418
595	MOTOR-ELECT OVR 20 H			2846	781	1743	322	197	163	324	254	476	389	236	230	214
599	MISC MACHINERY REPR			2226	633	614	981	216	512	117	190	806	174	119	603	1018
601	PIPING-PROCESS			6948	1473	5815	339	75	1291		232	1819	1554	188	577	418
603	PUMP-CENTRIFUGAL			0									1520			
607	PUMP-MOYNO-POS DSP			0								62	315	228		
610	PUMP-SUMP-PORT & STA			0												
632	SANDER			0							908			48		
647	TANK-METAL			0												31
674	VALVE-FLOW CONTROL			0												
	OPERATION TOTAL	4507	5366	23683	3785	10788	9112	893	3893	3631	9375	12206	7358	6931	7052	6517
	OPERATION 090															
	BOARD MILL GENERAL			256												
001	MACHINERY EXPENSE											1				
003	MACH REL/REM EXPENSE					254	2	203			235	63				
004	BUILDING EXPENSE									220					107	
007	LE & T EXPENSE							71	90							
009	*									4	44		24			
071	EXP-OSHA MACHINERY								281					39	235	19
200	AIR COND.-WINDOW														88	110
208	DOCK BOARDS														62	35

Figure 10.12 Section from a monthly summary report of maintenance performed at a specific plant.

ME515
AREA 2   OPERATION 830

PENSACOLA PLANT OPEN JOB REPORT
WEEK ENDING 01-31

JOB NUMBER	DATE ISSUED	DATE WANTED	PRI	DESCRIPTION	HOURS	ESTIMATED TOTAL $	ACTUAL HOURS	LABOR AND OVERHD $	MATRL $	OUTSIDE $	TOTAL $
EQUIPMENT CLASS 001											
002520807	03-07		5	MODIFY GRACO PUMP NO 1 PL	16.0	210	16.0	210			210 **
002527607	07-28		5	MODIFY PYROJET BURNERS	48.0	792					
002534207	11-03	10-31	4	RELOCATE RVS DR 830-748 CONV	16.0	224	14.0	179	8		187 *
002537207	12-05	11-29	4	MODIFY WATER LINE NO 1 BT NO 1	32.0	620	50.0	703	466		1169 **
004830007	01-16	11-22	4	MODIFY OPENINGS NO 1 BOOTH NO 1		323				323	323 **
005372801	11-06		5	USE 1 PL OVEN EXH IN COOL SECT		1600				1491	1491 *
005377101	12-05	02-01	4	FAB BRIDGE CONV FOR 1 PNT LINE	80	1420					
005377501	01-10	02-15	3	INSTALL BLOWOFF HDR ON 1 PL		1190					
064990002	09-06	03-01	4	IRONNG ROLLS FOR NO. 1 PAINT L	34.0	34700		15111	930		930 **
064996002	09-12	03-26	4	FABRICATE IRONING ROLLS		33641			21144		36255 **
EQUIPMENT CLASS 204											
005954307	09-11	08-28	3	REPR HOOK CHAIN FALL NO1PL IRON	2.0	33					
EQUIPMENT CLASS 523											
001682507	09-09	09-07	4	REPR NO1 PL PYROJET BRN BRK	4	56					
EQUIPMENT CLASS 536											
001705907	01-09		E	REPL BAD BRKR FOR SPEED FOIL HT	10.0		28.0	358			358 **
EQUIPMENT CLASS 537											
001164207	01-12	01-10	2	REPR CONDUIT SPEED FOIL HEATERS	6.0	124					
001992007	01-07	01 07	E	REPR ELECT CONT PANEL NO 1 PL							
005903707	01-12	01-10	2	REPR EL WIRE FEEDER CONTROL BOX	4.0	76	8.0	122			122 **
EQUIPMENT CLASS 542											
001400907	01-08		5	REPL BRG 830-8 CONV	16.0	340	14.0	198			198
EQUIPMENT CLASS 570											
001991907	01-06		E	REPL NO 1 PL E CIRC FAN BRG	4.8	72	4.8	72			72 **

Figure 10.13 Maintenance open job report for a specific plant as generated by a specific maintenance system.

PENSACOLA PLANT PRIORITY REPORT
RUN DATE 01-08-85

MAINTENANCE AREA 01	JOB ORDER COUNT	HOURS EXPENDED LAST MONTH	PERCENTAGE TOTAL HOURS
PRIORITY E	41	490	10.6
PRIORITY 1	25	295	6.4
PRIORITY 2	97	3019	65.4
PRIORITY 3	5	12	0.3
PRIORITY 4	41	778	16.8
TOTALS	209	4618	99.5

Figure 10.14 Computer-generated report showing distribution of maintenance jobs by priority level.

Reports should not be generated unless they are utilized for improvement purposes. A few reports that let you know how you are doing and where improvement can be made are all that are necessary.

SUMMARY

An equipment register, which is a computer data base created from a coding sheet, should be kept for all capital equipment. The equipment register that is maintained should carry information pertinent to the equipment location, manufacturer, and service characteristics; lubrication specifications; and repair summary. Space should also be available for estimating time and cost of anticipated maintenance.

The progressive company today utilizes five principal computer applications in connection with its maintenance management effort. These are in addition to those systems that handle payroll, labor accounting, and so on.

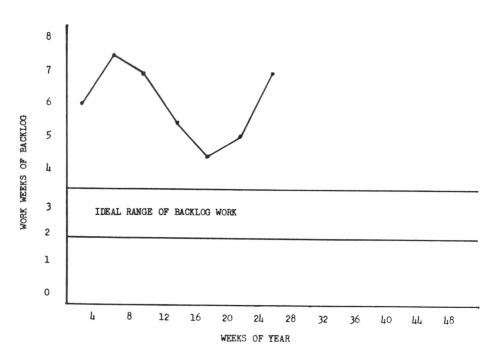

Figure 10.15 Work order backlog for 6 mo of millwright activity.

These five systems are

1. Equipment control—for all the facilities that the business operates that must be maintained and serviced. On occasion equipment will fail and must be brought back into service with a maintenance effort.
2. Work control—for the personnel that perform maintenance and repair work. From predetermined time standards, work is scheduled in an effective manner.
3. Maintenance spare parts and inventory control—for stock and stores used in maintaining, repairing, and overhauling equipment. Inventories are maintained in efficient quantities and are controlled in the most cost-effective manner.
4. Cost accumulation and reporting—for management control and planning. The cost of production facilities is accumulated and recorded so as to determine when to purchase new equipment and of what type.
5. Performance reporting—for management to make strategic decisions to improve performance of all equipment and increase its availability.

SELECTED BIBLIOGRAPHY

Finely, Howard F. *Modern Maintenance Management Systems*, Houston, TX: The Howard Finley Corp., 1981.

Nicks, J. E. *Basic Programming Solutions for Manufacturing*, Dearborn, MI: Society of Manufacturing Engineers, 1982.

Otis, I., and VanZile, Donald K. "Maintenance Use of Computers." In Gavriel Salvendy, Ed., *Handbook of Industrial Engineering*. New York: Wiley, 1992.

11
Maintainability, Reliability, and System Effectiveness

The vast majority of the successful designs have given considerable consideration to both reliability and maintainability. The degree to which these attributes are incorporated in a product determine the system effectiveness and customer satisfaction. The objective of designing for maintainability is to provide equipment and facilities that can be serviced efficiently and effectively and repaired efficiently and effectively if they should fail. Equipment should be designed with sufficient reliability so that it will be operable for an anticipated life cycle at optimum availability. Thus reliability is a function of design; once the design has been completed and released for manufacturing, the reliability of the product or system has been determined—it can not be altered without redesign. Functional designs, where the technology of maintainability has been given considerable consideration, will inevitably result in simplified maintenance that can be performed both effectively and inexpensively. Equipment will be down for service or repair for a considerably smaller proportion of the time, and the anticipated life of the equipment will be longer. Thus maintainability is a basic characteristic of equipment, and the need for maintainability is affected by its reliability characteristics. Since improving maintainability is both difficult and costly after development, it is extremely desirable that reliability re-

Table 11.1 Responsibility for Failure in Electronic Equipment

Cause of failure	Total failures (%)
Design	33
1. Electrical considerations	
a. Circuit and component deficiencies	11
b. Inadequate component	10
c. Circuit misapplication	12
2. Mechanical considerations	10
a. Design weaknesses, unsuitable materials	5
b. Unsatisfactory parts	5
Operation and maintenance	30
1. Abnormal or accidental condition	12
2. Manhandling	10
3. Faulty maintenance	8
Manufacturing	20
1. Faulty workmanship, inadequate inspection and process control	18
2. Defective raw materials	2
Other	7
1. Worn out, old age	4
2. Cause not determined	3

Source: NEL Reliability Design Handbook, U.S. Navy Electronics Laboratory, San Diego, CA.

quirements of facilities be included in design specifications and that designs give attention to maintainability.

One must recognize that no product can be assumed to have 100% reliability at any point in its life cycle—even in the first minutes of its use. However, successful designs should have 100% maintainability.

Most major maintenance problems are better solved at the equipment-design level rather than at the level of maintenance-personnel training (see Table 11.1). Maintainability problems solved at the design level are much less costly than teaching technicians to deal with the countless contingencies that may arise during the life of the equipment. Obsolescence and changes in successive production models of the same equipment may vitiate much of this prolonged training before it is completed.

Maintainability implies a built-in characteristic of the equipment design and installation which imparts to the cell an inherent ability to be maintained, so as to keep the equipment productively operating by employing a minimum number of maintenance man-hours, skill levels, and maintenance costs.

GENERAL OBJECTIVE OF MAINTAINABILITY

The principal objective of maintainability is to maximize the availability of all facilities and equipment utilized by an industry or business. The basic needs related to all equipment are performance, reliability, and maintainability (Fig. 11.1). All three of these needs must be met in order to provide the greatest availability. High performance from reliable equipment designed with poor maintainability will result in higher costs since this equipment does not include the design characteristics which speeds return of down equipment to a running state with the minimum requirements for time, resources, manpower, and technical skills.

RELIABILITY

Today, reliability requirements of equipment, machines, and electronic hardware and systems are generally included in design specifications. Also, reliability quantitative data is a property of facilities and systems included in

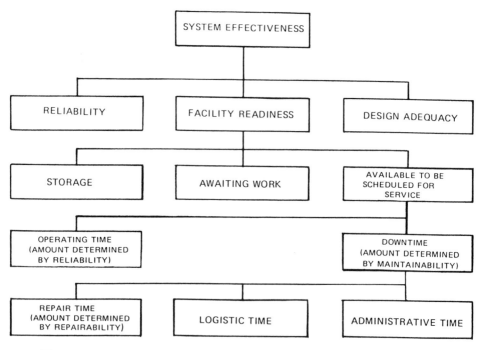

Figure 11.1 Relationship between reliability, maintainability, and system effectiveness.

military specifications. Reliability can be considered as being a characteristic of design which results in durability of the product or system while performing its intended use over a predetermined time interval. High reliability is achieved by proper selection of sound engineering principles, materials, sizing, manufacturing processes, inspection, testing, and total quality control.

Thus, we can define reliability as being the probability that a product or system will operate successfully under a specified environment for a certain time duration. It should be apparent the reliability characteristics of a product change with time. The reliability of a product can be expressed as

$$R(t) = e^{-\lambda t}$$

where

$R(t)$ = the reliability at any time t

e = the base of napierian logarithms = 2.303

λ = the total number of failures per operating period

t = planned operating time

The reliability function for exponential distribution appears in Fig. 11.2.

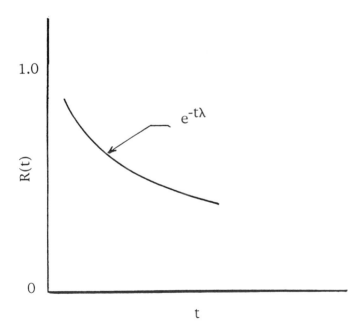

Figure 11.2 Reliability function for exponential distribution.

Estimating the Reliability of Equipment

If a piece of equipment represents the assembly of several components and each component has its own pattern of failure, then obviously the probability distribution of failure will be different for each component. The reliability of the assembly will be equal to the product of the separate reliabilities of the components going into the assembly. For example, if an assembly is made up of four components with the following reliabilities at 1000 hr of service:

$$R_1 = 0.95, \quad R_2 = 0.98, \quad R_3 = 0.97, \quad R_4 = 0.90$$

the reliability of the assembly at 1000 hr of service would be

$$
\begin{aligned}
R_a &= R_1 \times R_2 \times R_3 \times R_4 \\
&= (0.95)(0.98)(0.97)(0.90) \\
&= 0.813
\end{aligned}
$$

Therefore, we could expect about one of these assemblies out of every five to fail after about 1000 hr of service.

ANTICIPATING BREAKDOWNS

The majority of installed equipment is contained in the constant hazard portion of the mortality curve. Equipment in this period can be assumed to be in a random failure period where the failure rate, λ, is

$$\lambda = \frac{\text{total number of failures}}{\text{total operating hours}}$$

λ is estimated from historical data and can usually be provided by the supplier of the equipment. With predictive maintenance, it is necessary to know the length of time that can be expected between breakdowns. This period is referred to as MTBF (mean time between failures) and is equal to

$$\text{MTBF} = \frac{1}{\lambda}$$

In addition to knowing the MTBF, the analyst doing predictive maintenance will want to know the length of time that an unscheduled breakdown and repair will take. This time is comprised of two components: (a) the time spent waiting for spare parts or special materials and (b) the expected time required to make the actual repair. The sum of these two times can be identified as the mean forced outage time (MFOT) of the capital equipment.

The total average forced outage time per total operating hours would equal $\lambda \times$ MFOT.

For example, a given piece of equipment may involve six components subject to failure. The estimated number of failures per 1,000,000 hr of service (λ), mean forced outage time (MFOT), and their product are:

	Failure rate (λ)	MFOT (μ)	Product ($\lambda\mu$)
Drive belt	152	0.5	76
Bearings	25	24.0	600
Breaker and rel.	31	8.0	248
Transformer	14	72.0	1008
Fuse switch	9	1.0	9
Relay	27	8.0	216
Sum	258		2157

From the preceding, the analyst can estimate the mean time between failures of this piece of equipment to be

$$\text{MTBF} = 10^6/258 = 3876 \text{ hr}$$

Assuming this equipment operates 2 shifts a day (16 hr), 5 days/wk, we would anticipate that once in every 0.93 yr on the average, there will be a breakdown on this piece of capital equipment.

$$3876/(16)(5)(52) = 0.93 \text{ yr}$$

In this example, the mean forced outage time is computed by dividing the sum of the waited product component outage by the summation of the component failure rate.

$$\text{MFOT} = \frac{\Sigma\lambda\mu}{\Sigma\lambda} = \frac{2157}{258} = 8.36 \text{ hr}$$

The reader should understand that the component MFOT can be calculated by estimating the wait time to acquire a spare component from the inventory system in effect (see Chapter 6) and estimating the repair operation time from standard data or slotting techniques (see Chapter 4).

Once the analyst has computed the predictions of MTBF and MFOT, the probability of the work stoppage of the equipment and its availability can be determined. The probability of a breakdown or the unreliability of a piece of equipment provides a means for estimating the preventive maintenance which will be required.

$$Q = 1 - R$$
$$= 1 - e^{-t\lambda}$$

where

Q = unreliability

R = reliability

λ = failure rate

t = planned operating time

The time period that the predictive maintenance analyst is concerned with is the time between planned maintenance shutdowns. During planned maintenance shutdowns, worn-out parts are replaced, equipment is cleaned, lubricated, and adjusted so that the system will continue to remain in the random failure (constant hazard) period the rest of its life.

The system availability predicts the actual running or uptime in terms of the ratio of actual operating time to the scheduled operating time.

Thus

$$\text{Availability} = \frac{\text{actual operating time}}{\text{standard operating time}}$$
$$= \frac{\text{MTBF}}{\text{MTBF} + \text{MFOT}}$$
$$= 1 - \text{unavailability}$$

Based upon the previous example, the probability of a failure for a 1-year period, with a 3-day planned shutdown for overhaul (3 × 16 hr/day) would be computed as follows:

$$Q = 1 - e^{-t\lambda}$$
$$= 1 - e^{-(258 \times 10^{-6})(4160 - 48)}$$
$$= 1 - 0.346 = 0.654$$

This indicates that each facility of this type has a 65% chance of failure in the next year under two-shift, five day/wk operation.

The availability, A, for this piece of equipment is as follows:

$$A = \frac{\text{MTBF}}{\text{MTBF} + \text{MFOT}}$$
$$= \frac{3876}{3876 + 8.36}$$
$$= 0.9979$$

In order to convert this availability into expected downtime, the scheduled operating time is multiplied by the unavailability.

Predicted average downtime hours $= t(1 - A)$

$= [(52 \text{ wk} \times 5 \text{ days} \times 16 \text{ hr}) - 48 \text{ hr}](1 - 0.9979)$

$= 8.635 \text{ hr/yr}$

SYSTEM EFFECTIVENESS

The relationship between reliability, maintainability, and system effectiveness is illustrated in Fig. 11.2. It can be seen that system effectiveness is dependent upon availability which in turn is a function of both reliability and maintainability. Availability is the probability that a facility scheduled for service will be operating at any point in time. Thus availability can be expressed as

$$A = \frac{\text{MTBF}}{\text{MTBF} + \text{MTTR}}$$

where

MTBF = mean time between failures = $1/\lambda$

MTTR = mean time to repair

λ = total number of failures per total operating period

The reader should recognize that high reliability results in a high MTBF and good maintainability results in a small MTTR. Sound design provides both high reliability and maintainability.

An increase in maintainability or reliability or both will increase the availability of the system. So from a decision-making point of view, availability can be expressed as a function of maintainability and reliability as

$$A = f(R, M)$$

where

A = availability of the facility or system

R = reliability of the facility or system

M = maintainability of the facility or system

From a geometric concept, boundary limits for a given availability can be identified as relating to specific levels of maintainability and reliability (see Fig. 11.3). Note that for a given contour line, say 0.70, the trade-off for one parameter, say reliability gain, is not offset by an identical loss in maintainability.

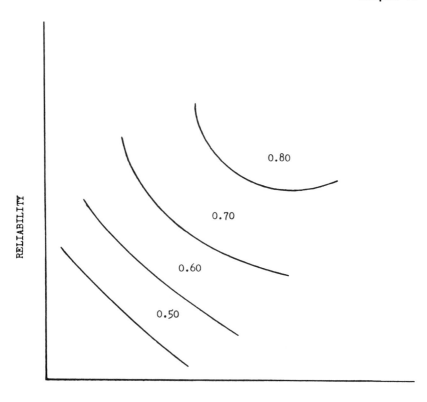

MAINTAINABILITY

Figure 11.3   Plots of availability ranging from 0.50 to 0.80.

GENERAL PRINCIPLES OF MAINTAINABILITY

The following fundamental principles should be considered by the functional designer in the development of components, products, and systems so that both reliability and maintainability are designed and built into the equipment:

1. Endeavor to eliminate or minimize the need for maintenance. This principle is basic and always is considered by the conscientious product designer. For example the following considerations may be evaluated:

    a. Would nylon (or other plastic) gears, which would not require maintenance, be suitable?

b. Can standard "sealed for life" bearings be utilized with the present economic constraints?

c. Can electrical components that are sealed and of adequate size, such as condensers, resistors, and transformers, be specified so as to assure freedom of failure?

2. Minimize the frequency and complexity of required maintenance tasks. For example, if a part of a design needs to be periodically serviced, tested, and calibrated; then this part should be able to be removed easily so that the maintenance functions can be performed readily. Thus, quick disconnects should be provided and the time needed to remove the component should be established. When component removal is not possible or practical, the design should provide entry ports so that the part can be reached easily and maintained with standard tools.

3. Endeavor to require maintenance that utilizes low skill levels and minimum training to perform the function adequately. Parts requiring maintenance should not only be accessible, but they should also be able to be maintained using standard tools requiring low skill levels and physical and mental effort. The maintenance function should be clearly spelled out so that the mechanic is not required to execute a high level of initiative or judgment.

4. Determine the optimum preventive maintenance that should be performed. As already brought out (Chapter 8), planned maintenance is superior in the vast majority of cases to breakdown maintenance. In the product design phase, the ideal amount of preventive maintenance should be established.

5. Establish complete information to be used in the education and training for maintenance. At times unique maintenance will need to be performed on certain components or systems incorporated in the design of the equipment. Details of how, when, how much, and why regarding the maintenance of all equipment should be established so that maintenance personnel can be trained to perform the required maintenance utilizing the best methods, assuring the quality of the maintenance. The material developed should be concise, simple, and easily understood. Brevity is encouraged. The ample use of drawings and sketches is encouraged so that the recommended maintenance procedure is understood. Procedures should be developed to insure that the simple, adequate technical data are available with the equipment when delivered.

6. Endeavor to provide components that can be adjusted for wear. Develop the design so this adjustment can be performed without tear-

down. The ability to tune an engine by adjusting the ports of the carburetor, the points of the distributor, and the gaps of the spark plugs is an example of cost-effective maintenance that is superior to simply replacing these parts.

7. By utilizing standard data (see Chapter 4), develop time standards for the removal, installation, and repair of major items of the equipment that will probably need to be serviced during the product's life cycle. If this point is observed, maintainability in newly designed equipment is practically assured. When the designer is charged with the responsibility of specifying how long servicing tasks should take, knowing this time should be minimized, an effort will be made to develop a design that allows these operations to be performed both easily and rapidly.

8. Provide optimum accessibility to all equipment and components requiring frequent maintenance, inspection, removal, replacement, or adjustment. This fundamental principle, although easily understood, is often violated. Every experienced maintenance employee can identify many maintenance functions that had to be performed where the part being serviced was hidden in such a manner that it was extremely difficult to service or remove. It is in the design stage that accessibility can be provided. Sometimes this can be achieved by adding a window or port through which the component can be reached and serviced. At other times, it may be advantageous to remove an entire subassembly in order to service a specific component. For example, engines and motors should be able to be replaced rapidly. The functional designer should provide the human engineering aspects for access to facilitate electrical, pneumatic, hydraulic, lubrication, and fuel servicing. Provisions should be made for the ease of inspection, replacement, and rapid adjustment in servicing of brakes, clutches, and batteries without the need of teardown. Any component requiring frequent maintenance should be located to preclude the need to remove other components to gain access to the specific component. All lubrication plugs and fittings should be readily accessible.

9. In the design of equipment and facilities, provide for rapid and positive identification of equipment malfunction or marginal performance. This may include troubleshooting charts in fault-tree diagram form from which potential failures and the steps to correct them can be identified. Frequently, instrumentation can identify impending trouble. For example, in the falloff of oil pressure an idiot light can be activated or a discharging situation can result in activating an auditory sensor.

10. It should be possible to determine the performance of the facility readily so that adjustments or repairs can be anticipated. Speed, power, surface finish, vibration, and noise are typical parameters that usually are able to be measured in order to assure normal performance of equipment. The provision of test points for accepting automatic test equipment is frequently desirable.

11. All parts and components should be able to be positively identified so that they can be repaired readily or replaced. Color coding is frequently helpful in addition to part numbering. Identification as to component rating, type of fuel, lubricant, liquid, and gas utilized should be clearly visible after extensive use.

12. Assemblies should be designed to avoid the necessity of special tools and fixtures in their disassembly and reassembly. A set of standard tools should be identified that can be used to maintain efficiently the whole commodity. Provision for quick-disconnect devices for rapid removal and assembly of components is advisable.

13. In the design of equipment avoid the exclusive use of proprietary items and critical, costly, rare, or difficult to process materials. Components and materials such as these may be difficult to inventory in adequate quantity and can lead to costly repair and overhaul delays.

14. Always design to provide for maximum interchangeability. Interchangeability inevitably simplifies maintenance, reduces parts inventory, and allows for more economical costs of the equipment and its operation.

15. In heavy components, provide for the facilitation of transport such as towing, hoisting, lifting, and jacking. A strategically placed jack support so that a part can be elevated easily by jacking can save considerable time during the repair of a part and often will prevent a costly or serious accident.

16. Provide bearings and seals that require the minimum of replacement and servicing on a life cycle basis.

17. Specify gears that will satisfy all anticipated overload requirements. Likewise, other mechanical, electronic, electrical, hydraulic, and structural components should be of such capacity as to combat overloads.

18. Endeavor to combine components of a system or function of a system into a removable assembly.

19. Provide for self-adjustment of components subject to wear; for example, tapered roller bearings and end seals as opposed to radial seals.

20. On instruments and other components that are easily damaged by shock and vibration, provide for isolation to avoid such conditions.

Bonded rubber shear-type mountings are usually effective vibration isolators.

21. Specify corrosion resistant materials and protective coatings, plating, and metallizing. Avoid placing bare dissimilar metals in contact with each other, thus preventing galvanic corrosion. Utilize wire insulation in coils that will withstand high temperatures. Equipment reliability is always enhanced by utilizing superior materials.

The preceding 21 general principles can be expanded. However, they provide basic guidelines to be observed in the design and development of new equipment so that maintainability is incorporated in the design to a high degree.

SPECIFIC DESIGN CONSIDERATIONS

Maintenance Access

The functional design should always provide an access wherever a frequent maintenance operation would otherwise require removing a cover or housing, opening a connection, or dismantling a component. The access should be designed of sufficient size and shape so that the repair or servicing operation is easy to perform. Typical accesses include entrance doors; inspection windows; and lubrication, pneumatic, and hydraulic servicing points.

The location of the access is largely based on how the equipment will be installed. All accesses, displays, controls, and so on, ideally should be on the same face of the equipment since it frequently is difficult to keep more than one face accessible. An effort should be made to keep accesses away from nip points, moving parts, high voltages, and other dangerous areas so that the chance of injury to the maintenance mechanic is minimized. Table 11.2 outlines recommended equipment accesses.

Tool Requirements for Maintainability

The functional design should permit normal maintenance using standard tools. The use of special tools is justified only when a common tool cannot be utilized or when a special tool will facilitate a maintenance task in terms of reduced time or increased accuracy.

Only about 5% of all hand tools are extensively used by maintenance mechanics. This includes screwdrivers, pliers, wrenches, and soldering irons. Most designs specify a wider variety of fastener (nuts, bolts, screws, etc.) sizes than necessary. From a maintenance standpoint, it would be ideal to

Table 11.2 Equipment Accesses

Desirability	For physical access	For visual inspection	For testing and service
Most desirable	Pullout shelves or drawers	Opening with no cover	Opening with no cover
Desirable	Hinged door (if dirt, moisture, or other foreign materials must be kept out)	Plastic window (if dirt, moisture, or other foreign materials must be out)	Spring-loaded sliding cap (if dirt, moisture, or other foreign materials must be kept out)
Less desirable	Removable panel with captive, quick-opening fasteners (if there is not enough room for hinged door)	Break-resistant glass (if plastic will not stand up under physical wear or contact with solvents)	Screw-on cap with few large pitch threads (if dirt, moisture, or other foreign materials must be kept out)
Least desirable	Removable panel with smallest number of largest screws that will meet requirements (if needed for stress, pressure, or safety reasons)	Cover plate with smallest number of largest screws that will meet requirements (if needed for stress, pressure, or safety reasons)	Cover plate with smallest number of largest screws that will meet requirements (if needed for stress, pressure, or safety reasons)

specify but one size of screw, bolt, and nut. Then only one size of screwdriver and wrench would be needed to perform normal maintenance functions. This is, of course, seldom practical, but the point that functional designers need to consider is the many advantages of standardization by limiting the variety and size of fasteners.

Equipment should be designed so that straight screwdrivers can be used to set and remove the fastener (see Fig. 11.4).

The functional designer should endeavor to develop designs that do not require the need for torque wrenches. This specialized tool may be overlooked by the typical technician when reassembling an assembly. Another point that should be considered is the amount of clearance provided for wrenches. There must be enough space available for the tool to operate after being positioned on the nut or bolt head. This clearance should allow the use of an adjustable wrench which will require more clearance than a fixed-size wrench.

In connection with screwdrivers used by the maintenance mechanic, the following points should be considered:

1. Screwdrivers for small adjustment screws should be equipped with a funnellike pilot that will guide the bit into the screw slot (see Fig. 11.5a).
2. Magnetized screwdrivers or ones with clips should be used to hold small screws in position when it is difficult to position the screws with one's fingers (see Fig. 11.5b).

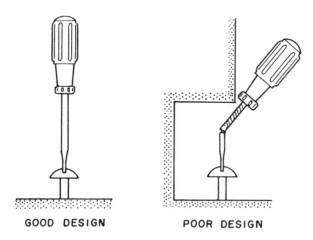

GOOD DESIGN   POOR DESIGN

Figure 11.4 Good design allows the use of straight screwdrivers.

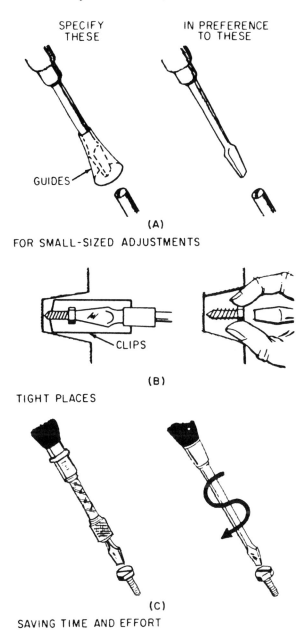

SPECIFY
THESE

IN PREFERENCE
TO THESE

GUIDES

(A)

FOR SMALL-SIZED ADJUSTMENTS

CLIPS

(B)

TIGHT PLACES

(C)

SAVING TIME AND EFFORT

Figure 11.5 Screwdrivers for various purposes.

3. Push-type screwdrivers should be used for removing and driving screws that must be rotated for several revolutions (see Fig. 11.5c).

Identification

It is important that the maintenance technician be able to identify readily components, parts, controls, and test points in order to perform an assignment thoroughly, rapidly, and free from error. By proper identification through the adequate marking or labeling of parts, components, controls, and test points; the functional designer will facilitate repair and replacement during maintenance operations.

The information that usually is included with equipment is its electrical requirements, horsepower, rpm, weight, capacity, limits, and ranges. This information should be supplemented with information that describes and illustrates the basic operating instructions, wiring, fluid flow, calibration and adjustment instructions, safety precautions, location of test points, valve settings, ignition settings, types of fuel recommended, oils and greases preferred, and similar data.

Mechanical parts should carry identifying numbers. Thus such components as housings, brackets, pulleys, levers, bearings, and so on, should be appropriately marked with an identifying number during manufacturing.

Electrical items such as fuses and resistors should show their electrical rating in addition to an identifying number. The following points apply to all marking:

1. Be sure markings accurately identify the part.
2. Markings should be located either on the part or, if this is not practical, the markings should be adjacent to the part so there is no question as to what part the marking applies.
3. The markings should be permanent enough to provide certain identification for the life of the equipment.
4. Markings should be placed so that they are visible.

Interchangeability

It should be apparent that interchangeability has a positive impact on the ease of maintenance and repair. In equipment where frequent servicing and replacement of component parts takes place because of wear or damage, interchangeability should be practiced. Where interchangeability is practiced, it usually is necessary to have liberal tolerances. Thus those parts in equipment that will need to be replaced should be designed with the minimum number of sizes, types, assemblies, and subassemblies. Like assemblies, subassemblies and replaceable parts should be electrically, mechanically, hydraulically, or otherwise interchangeable both physically and functionally without electrical or physical modifications.

All components having the same manufacturer's part number should be directly and completely interchangeable with each other with respect to both performance and installation.

The following questions may provide encouragement to the functional designer to incorporate more interchangeability:

1. Can differences in size, style, shape, and so on, be avoided if they do not reflect different functional requirements in the design?
2. Has complete interchangeability been provided for those items that are intended to serve the same function?
3. Have mounting holes, slots, and brackets been designed so as to accommodate units (motors, for example) of different manufacturers?
4. Have all fasteners, connectors, lines, cables, and so on, been standardized throughout the system and from unit to unit within the system?
5. Are bolts, screws, and nuts of the same size for all covers and cases on a given design?
6. When complete interchangeability is not possible, can parts be designed for functional interchangeability with the use of adapters?

Hazards and Safety

Safety precautions should be designed into equipment to protect both the equipment operator and the maintenance mechanics who service the facility. If either the operator or the mechanic finds it necessary to divert attention from the work assignment in order to observe safety precautions, the job will probably not be done well.

Electrical shock is perhaps the major contingency that should be avoided by sound design for maintainability. Potentials ranging from 100 to 500 volts can cause death upon contact. Often severe injuries, resulting from reflex action causing the body to impact moving equipment, are a result of electric shock.

There are several safety procedures that will provide adequate personnel protection against electric shock. These include the enclosure of components that are the source of the power and using access-door safety switches, which can automatically open the main equipment switch as well as ground the components while the unit is open for access.

The safety switches that can be used to prevent electric shock include interlock switches. These switches will automatically open the power circuit when the access door is opened. On occasion equipment must be worked on with the power on, so there should be a switch mounted inside the equipment that permits maintenance workers to bypass the interlock system. After the access door is reclosed, interlock protection is restored.

Battle-short switches can be used to make interlocks inoperative. These switches are designed primarily for emergency use. The circuit involves a single switch that is wired in parallel with the interlock system. By closing the battle-short switch, a short circuit across the interlock switch is induced thus assuring power if there were an accidental opening of the interlock switch.

All electrically powered equipment should be furnished with a clearly identified main power switch. These switches should be equipped with adequate safeguards against large arcing. Also one should not be able to open the switch box with the circuit closed.

It is important to assure that adequate grounding techniques are incorporated to protect personnel from dangerous voltages. Machine chasis, equipment enclosures, and exposed parts should be maintained at ground potential using the same ground. A terminal properly secured by either spot welding or using a firm mechanical fastener provides a satisfactory ground connector. Good design suggests that the external ground conductor be produced from a flexible copper strap capable of carrying at least twice the current needed for the equipment.

Potential mechanical hazards, too, need to be considered in the design of equipment. Moving parts should be shielded by guards so that personnel are protected from accidentally contacting rotating or oscillating parts. Also, high-temperature parts should be guarded or located so that direct contact will not be made by workmen.

To prevent abrasions, cuts, contusions, and similar injuries, the equipment should be free from sharp edges and corners, thin edges, and protrusions from surfaces.

The following safety checklist should be considered in the design of new equipment.

1. Keep high (dangerous) voltages away from internal controls such as switches, adjustment screws, and commonly replaced parts.
2. Keep all live wires that retain voltages when the equipment is off in a position where they will not likely be contacted in normal service of the machine.
3. Provide conspicuous labels adjacent to both high-voltage and hot areas.
4. Provide grounding of electrically neutral parts of electrical systems.
5. Do not combine control and warning circuits.

6. Utilize color code techniques that clearly define operating and danger ranges.
7. Provide either a transparent window or removable cover over fuses so that they can be checked without removing the component case.
8. Provide guards on all moving parts of power transmission and machinery.
9. Protect access openings with rounded surfaces or utilize soft covers (rubber, plastic, etc.) over sharp edges to prevent injury.

IDENTIFICATION

With adequate identification of facilities, components, controls, test points, and so on, the problem with sound maintainability is simplified. All equipment should be identified with a securely attached nameplate. The nameplate lettering may be produced by any permanent marking procedure, such as engraving or stamping, and should provide such pertinent information as manufacturer's name and address, serial number, capacity, current requirements, horsepower, frequency, limits, ranges, weight, and other basic information.

Instruction plates of the permanent type should be attached to each facility in a conspicuous place so that they can be read easily. These instructions should describe and illustrate (where appropriate) wiring and fluid flow diagrams, operating instructions, adjustment instructions, location of test points, pertinent electronic equipment, recommended oils and lubricants, pressure ranges, ignition settings, valve clearances, and so on.

Cautions and warnings should be included to warn personnel of any hazardous conditions and should include precautions to be observed to ensure the safety and well-being of personnel and equipment.

The following checklist can help assure that adequate identification has been provided:

1. Is all equipment labeled with complete identifying data?
2. Are components identified with important characteristics?
3. Are structural or supporting members supplied with composition data indicating limitations such as ability to withstand elevated temperature? Can the component be subjected to welding?
4. Are terminals identified with the proper code of attaching wires?
5. Have identification labels and instructions been placed so they can be read easily?
6. Where color coding is used, has the meaning of the different colors been shown?

7. Is the color coding consistent in all applications?
8. Are schematics and instructions clearly presented and displayed so they can be used readily for troubleshooting purposes?
9. Are all lubrication points clearly identifiable and accessible?
10. Is important information concerning electrical, hydraulic, and pneumatic characteristics of components made available on display labels of component covers?

MAINTAINABILITY SPECIFICATIONS

Maintainability should be defined in specific terms. Both what and how the work should be done should be spelled out, along with what the crew size should be and how long it should take (MTTR) to perform the repair or maintenance.

Maintainability specifications are usually recorded in maintenance manuals, which can be thought of as being the communication link between the design engineer and the maintenance technician. The maintenance manual will provide instructions on the maintenance of equipment, facilities, and materials that are housed in the plant. These instructions should be written simply and should not require the technician to make mathematical calculations. In the maintenance function, the technician should not be required to consolidate or integrate information from different sources; collect, process, or report any unusual or complex data; post data from one from to another; or keep permanent records (see Fig. 11.6).

SUMMARY

The vast majority of product design engineering personnel are concerned first for high-quality design function, second for design cost (i.e., produceability), third for design appearance (aesthetics), and last for maintainability. Since these four qualities usually oppose one another (for example, an increase in maintainability may result in an increase in product cost of manufacture), that parameter that carries the lowest priority (fourth) will usually receive the least attention. Thus maintainability is usually sacrificed in competitively priced facilities such as machine tools and other capital equipment.

Designers today are beginning to realize that a quality product that is functionally sound is not adequate in our competitive society. In addition, the purchaser of capital equipment today expects both reliability and maintainability to be designed into items to insure the desired performance for their entire anticipated life cycles.

Maintainability Standard Practice Procedure
Assembly Machine #1140654

Assignment: Remove and replace propeller shaft #1140654-11

Period: 12 months of operation

MTTR: 3.75 hours   Crew size: 1

Tools: Kit 11-14

Operation:

1. Disconnect rear U-joint from drive pinion flange.
2. Pull drive shaft toward rear of the assembly machine until front U-joint yoke clears transmission housing and output shaft.
3. Install tool 11-14-3 in seal to prevent lube from leaking from transmission.
4. Before installing, check U-joints for freedom of movement. If a bind has resulted from misalignment after overhauling the U-joints, tap the ears of the drive shaft sharply with rubber hammer to relieve the bind.
5. If rubber seal installed on end of transmission extension housing is damaged, install a new seal.
6. Lubricate yoke spline with grease (TR-05).
7. Install yoke on transmission output shaft.
8. Install U-bolts and nuts which attach U-joint to pinion flange. Tighten U-bolts evenly to prevent binding U-joint bearings.

Figure 11.6 Maintainability standard practice procedure for replacing propeller shaft in a specialized assemby machine.

Consumers today expect that down equipment needs only simplified test and repair procedures in order to minimize the time required to locate and correct faults by providing ease of access and simplification of adjustments and repairs. It is important that engineering design personnel verify that maintainability has been incorporated in their work before they sign off their design assignment. Likewise, all experimental designs should be checked for maintainability just as they are checked for function and reliability. Opportunities for creativity in design are applicable not only in soundness from a quality function standpoint, but also from the standpoints of maintainability and produceability. This creativeness should be rewarded through overall evaluation of design soundness, that is, designing for quality performance, produceability, appearance, and maintainability.

SELECTED BIBLIOGRAPHY

Harring, Michael G., and Greenman, *Lyle R. Maintainability Engineering* (prepared under contract #DA-31-124-ARO-D-100-34). Martin Marietta Corporation and Duke University, 1965.

Nakajima, Seiichi, Yamashina, Hatime, Kumagai, Chitoku, and Toyota, Toshio. "Maintenance Management and Control." In Garriel Salvendy, Ed., *Handbook of Industrial Engineering,* 2nd ed. New York: Wiley, 1992.

Reliability Analysis Center. *Reliability Design Handbook.* RDG-376, Griffiss Air Force Base. New York, 1964.

U.S. Army Material Command. *Engineering Design Handbook—Maintainability Guide for Design.* Alexandria, VA: October, 1972

U.S. Army Material Command. *Engineering Design Handbook: Part 3, Reliability Prediction,* AMC Pamphlet No. 706-197. Alexandria, VA, 1976.

12
Utilities Management

Utilities management is mainly concerned with the control of costs at a given facility. Also, those in charge of utilities management are concerned with the conservation of energy and the elimination or reduction of harmful emissions that may have a detrimental effect on the environment. As brought out in Chapter 1, the conservation of energy is one of the four principal activities carried out by a well-organized engineered maintenance management system. Plant engineering has the responsibility for the design of a system to control the use of energy for operating machinery, heating, ventilation, air conditioning, hot water, and lighting of the various segments of the plant. The monitoring and maintenance of the control system is the responsibility of the maintenance function, which, of course, reports directly to plant engineering. In addition to monitoring the energy users and the control system, maintenance will need to inspect the building envelopes periodically including the roofs, windows, insulation, and so on, in order to maximize energy conservation.

Usually the largest utility cost is electricity, used for lighting, air conditioning, ventilation, and some space heating. However, the costs for natural gas and oil (for such processes as heat treating, forging, casting, and space heating) and water (for hydraulic testing of piping and vessels, drinking, cleaning, making of concrete) can be formidable. On an average, the cost of utilities when prorated over the total output of the plant is about 10% of the total cost of operation. This figure is only an approximation as the total utility cost can be substantially higher in connection with those processes using energy to transform a material, or lower where materials are utilized in their present form to produce a product. Today, there is usually at least a 30% improvement potential in the energy efficiency of a representative industrial facility.

With the sharp increases in electrical and other energy costs over the past few years, both large and small industries, businesses, schools, and hospitals, need to become more concerned about reducing or controlling their energy use. Energy costs are likely to continue to escalate and proper energy management will become even more important in the future than it has been in the past.

OBJECTIVES

The principal objective of the maintenance manager in connection with utilities management is to have the company use energy as efficiently as possible on all production and nonproduction operations being utilized throughout the plant. In order to realize this stated objective, the maintenance manager should have ongoing programs in

1. Utility usage and conservation. This will include understanding rate schedules and utility offered services such as cost-recovery studies, load analysis records, energy audits, and power quality studies.
2. Modern maintenance procedures throughout the plant building structure and equipment including heating and cooling systems.
3. Utility operation optimization.
4. Assurance of energy efficient designs of equipment and processes.
5. Active retrofitting based upon energy surveys and process changes and heat recovery techniques.

Typical projects resulting from these ongoing programs include the use of high-efficiency motors, heat recovery systems, variable-speed drives, combustion improvements, thermal insulation, cogeneration, waste steam utilization, and so on.

IMPLEMENTATION OF OBJECTIVES

Today in American industry the use of electricity is distributed approximately as follows:

Pumps	24%
Fans and blowers	14%
Compressors	12%
Machine tools	8%
Other motors	12%
DC drives	8%
HVAC	2%
Non-motor use	20%

The first step in achieving the objectives of the utilities management program is to collect data on how energy is being used in the plant. A study of this data along with a questioning attitude will bring out potential improvement possibilities. These improvement ideas will then be analyzed to determine their cost-effectiveness. Those that indicate a good potential will be implemented and followed up to determine if the estimated savings are being realized.

The initiation of the utilities management program will usually bring to light several places where energy is being wasted and where improvements can be made immediately with little if any capital investment. For example, an inspection that is focused upon energy waste will invariably reveal lights that should be turned off, light fixtures that should be cleaned, broken windows that should be replaced, HVAC filters that should be cleaned or replaced, cracks and holes that should be filled, leaking valves that should be repaired, and so on.

In the data collection procedure, the utility bill is useful. The rate structures from the electric, gas, oil, and water suppliers should be determined and analyzed for, at least, the past year. This will provide information on seasonal as well as production trends in connection with utility utilization. If there is any question regarding the bill, the utility should be contacted so that a representative can explain it.

The user can reduce the electric bill by knowing the rate schedule and (1) reducing the number of kilowatt-hours used and (2) lowering the peak demand kilowatt charges (some rates have a discount on off peak hours). The number of kilowatt-hours can usually be reduced by adding insulation to the ceiling and walls, replacing motors with high efficiency design (energy efficient AC motors provide 2 to 4% more efficient operation than standard motors), replacing standard incandescent lights with fluorescent bulbs (these use about 75% less energy), installing occupancy sensors to turn off lights in areas when people leave the area, replacing both electric heating and cooling equipment with more efficient models, using variable speed drives, and so on.

Also, in many instances, heat can be recovered from refrigeration units, air compressors, and heating furnaces to heat building areas or preheat water for steam production.

These improvements will also lead to lowering the peak demand charge. Peak demand charges can also be reduced by avoiding the same on-times of heavy loads.

ENERGY UTILIZATION INDEX

The energy utilization index (EUI) represents the energy (Btu) used per year per square foot of conditioned space. The EUI is a much used measure of a facility's energy performance. For example, if a building had 125,000 ft^2 of floor space and uses approximately 2,020,000 kWh and 4,800,000 ft^3 of natural gas in 1 yr, its EUI would be computed as follows:

$$3412 \text{ Btu/kWh} \times 2,020,000 \text{ kWh} + 1000 \text{ Btu/ft}^3 \times 4,800,000 \text{ ft}^3$$
$$= 6.892 \times 10^9 + 4.8 \times 10^9 \text{ Btu}$$
$$= 1.1692 \times 10^{10} \text{ Btu}$$

$$\text{EUI} = \frac{1.1692 \times 10^{10}}{1.25 \times 10^5} = 116,920 \text{ Btu/ft}^2/\text{yr}$$

LIGHTING

Modern fluorescent, metal halide, or high-pressure sodium lamp systems can result in significant power savings over conventional incandescent lamps at no sacrifice in the illumination level (see Fig. 12.1 and Table 12.1). The principal criteria that govern the visual environment are the amount of lighting, the contrasts between the immediate surroundings, the specific task being carried out, and the presence or absence of glare. The maintenance engineer should recognize that the ability to see increases as the logarithm of the illumination, so that a point is soon reached at which large increases in illumination will result in very small increases in the worker's efficiency.

As a general guide to the analyst, light recommendations in foot-lamberts are shown in Table 12.2, and footcandle requirements for different seeing tasks where the reflectance and contrast are relatively high are shown in Table 12.3.

A review of light requirements and the ability to achieve these levels through favorable location of the work station utilizing daylight, increasing

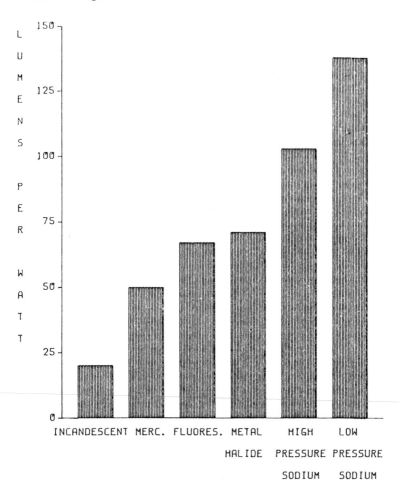

Figure 12.1 Luminous efficacy including ballast losses.

the percent of reflected light, and introducing modern lamps can result in dramatic power savings.

When reviewing the existing lighting system, the analyst will need to know the type of lighting already installed, what its connecting load is, how much lighting is activated during the various shifts of the working day, for how many days per year, how much lighting energy is consumed each year, and at what cost. Also, he will want to know how much light the present system actually is providing and what is its quality. Information should be acquired

Table 12.1 Representative Lamp Performance Data

Type	Watts	Color	Application	Initial lumens	Life (hr)	Lumens per watt
Low-pressure sodium	180	Yellow	Large areas	33,000	18,000	145
	135	Yellow	Large areas	22,500	18,000	121
	90	Yellow	Large areas	13,500	18,000	100
	55	Yellow	Large areas	8,000	18,000	96
	35	Yellow	Large areas	4,800	18,000	78
High-pressure sodium	1,000	Golden white	Large areas	140,000	15,000	127
	400	Golden white	Outdoor-indoor	50,000	20,000	111
	360	Golden white	Outdoor-indoor	34,200	12,000	83
	250	Golden white	Outdoor-indoor	25,500	15,000	88
	150	Golden white	Outdoor-indoor	16,000	12,000	107
	100	Golden white	Outdoor-indoor	9,500	12,000	79
Metal halide	1,500	Bluish to white	Sports lighting	155,000	1,500	95
	1,000	Bluish to white	Indoor only	88,000	10,000	82
	1,000	Near white	Enclosed fixt.	100,000	10,000	92
	400	Bluish to white	Indoor only	34,000	15,000	76
	175	Bluish to white	Enclosed fixt.	14,000	7,500	69
Deluxe mercury vapor	1,000	White	Outdoor-indoor	63,000	24,000	59
	400	White	Outdoor-indoor	22,500	24,000	50
	250	White	Outdoor-indoor	12,100	24,000	43
	175	White	Outdoor-indoor	8,150	24,000	41
	100	White	Outdoor-indoor	2,850	24,000	24

	Watts	Color	Application	Lumens	Life (hr)	CRI
Clear mercury vapor	1,000	Bluish	Outdoor	57,000	24,000	53
	400	Bluish	Outdoor	21,000	24,000	47
	250	Bluish	Outdoor	11,200	24,000	40
Self-ballasted	750	Near white	Indoor	14,000	16,000	19
mercury vapor	450	Near white	Indoor	9,500	16,000	21
	300	Near white	Indoor	7,800	20,000	26
	160	Near white	Indoor	2,700	20,000	17
Fluorescent	215	Standard	Outdoor-indoor	14,500	10,000	62
	215	Deluxe	Outdoor-indoor	10,000	10,000	43
	115	Standard	Outdoor-indoor	6,750	9,000	50
	115	Deluxe	Outdoor-indoor	5,000	9,000	37
	110	Standard	Outdoor-indoor	9,200	12,000	67
	110	Deluxe	Outdoor-indoor	6,600	12,000	49
	60	Standard	Outdoor-indoor	4,300	12,000	54
	60	Deluxe	Outdoor-indoor	3,050	12,000	38
	75	Standard	Outdoor-indoor	6,300	12,000	65
	75	Deluxe	Outdoor-indoor	4,500	12,000	46
	40	Standard	Outdoor-indoor	3,150	20,000	61
	40	Deluxe	Outdoor-indoor	2,200	20,000	42
Incandescent	1,500	Near white	Sports lighting	34,400	1,000	23
	1,000	Near white	Sports lighting	23,740	1,000	24
	750	Near white	Indoor-outdoor	17,040	1,000	23
	500	Near white	Indoor-outdoor	10,850	1,000	22
	300	Near white	Indoor-outdoor	6,360	750	21
	100	Near white	Indoor-outdoor	1,740	750	17

Table 12.2 Guide Brightness for Various Categories of Seeing at Specified Footcandles and Reflectance

Category of seeing task	Guide brightness (footlamberts)	Footcandles for specified reflectance conditions		
		90%	50%	10%
A. Easy	Below 18	Below 20	Below 36	Below 180
B. Ordinary	18–42	20–95	36–84	180–420
C. Difficult	42–100	45–153	82–240	420–1,200
D. Very difficult	120–420	133–445	240–840	1,200–4,200
E. Most difficult	420 up	455	840	4,200

Source: *Human Factors Engineering*, 4th ed., by Ernest J. McCormick (New York: McGraw-Hill, 1976).

Table 12.3 Current Light Recommendations for Different Seeing Tasks

Tasks	Footcandles in service, i.e., on task or 30 in. above floor
Most difficult seeing tasks	200–1,000
Finest precision work involving Fine detail, poor contrasts, long periods of time, extrafine assembly, precision grading, extrafine finishing	
Very difficult seeing tasks	100
Precision work involving fine detail, fair contrasts, long periods of time, fine assembly, high-speed work, fine finishing	
Difficult and critical seeing tasks	50
Prolonged work involving fine detail, moderate contrasts, long periods of time, ordinary benchwork and assembly, machine shop work, finishing of medium-to-fine parts, office work	
Ordinary seeing tasks	30
Involving moderately fine detail, normal contrasts, intermittent periods of time, automatic machine operation, rough grinding, garage work areas, switchboards, continuous processes, conference and file rooms, packing and shipping	
Casual seeing tasks	10
Stairways, reception rooms, washrooms and other service areas, active storage	
Rough seeing tasks	5
Hallways, corridors, passageways, inactive storage	

Source: International Labour Office, Geneva, Switzerland.

260

as to the type of control mechanism installed and the maintenance that the control system requires.

A modern lamp installation usually will not only save power, but will result in lower costs for replacement lamps and lamp replacement labor. Other important benefits—increased quality productivity, less absenteeism, and better worker morale—are likely to result as the new lighting system more closely matches footcandle requirements for differing seeing tasks throughout the company or business.

Of course, any energy management program includes education of the entire work force in connection with power costs. People must develop the habit of turning off lights when they leave a room or work area as well as any power equipment that is not needed immediately.

Programmable Lighting and Temperature Control

Lighting, heating, and air conditioning can be effectively scheduled throughout a facility using a microprocessor-based energy management controller. Controllers today are able to provide operating instructions to several thousand control points. The typical controller directs the operation of 32-point transceivers which receive and decode instructions and turn 24-v latching relays on or off. The relays can control the lighting by being inserted in series with circuit breakers on lighting panels as well as other electrical loads such as heating, ventilating, and air conditioning.

For example, a lighting program taking advantage of daylight can call for the controller to turn on warehouse lights in the month of January at the 50% level between the hours of 6:30 A.M. to 5:00 P.M. and at 100% level between 5:00 P.M. and 6:30 A.M. As the period of daylight increases, so will the lighting schedule change in many areas throughout a plant or business, including outside parking areas.

Under the programmable control; lighting, heating, ventilating, and air conditioning are programmed to turn off automatically. Minimum lighting and services will be provided for those periods such as weekend or night shifts when 100% lighting and service is not necessary. In the well-designed programmable system, local switches are available in convenient locations. Thus, someone who is working on a weekend when the programmable system has turned off the lights in the immediate area has only to push a button on a centrally located panel that will override the control and turn on the needed lights for a limited time.

Figure 12.2 illustrates a block diagram of a typical control system that links the computer with power circuits throughout a plant or business. The

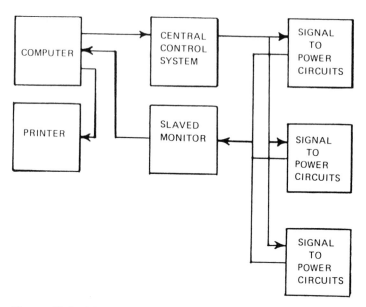

Figure 12.2 Block diagram of a typical control system.

system is able to control a large number of distinct channels or control points (although the illustration shows but three). There is really no limit on the number of points that can be controlled.

Electronic Infrared Sensor Controls Lighting

Electronic infrared sensors that operate by sensing and responding to the presence of the infrared heat generated by the human body can be used to control lighting. For example, such sensors can be used to automatically turn lights on when people enter a specific area of the plant or business and off when they leave. Typically such sensors are mounted in the ceiling of an area. This control will always turn lights off when people leave an area—a situation that frequently does not occur when this action is handled manually.

BOILER/HEATING SYSTEM

The boiler/heating system of most industries, institutions, and businesses is an important area to be audited in order to conserve energy. It is not unusual for a poorly maintained boiler to contribute up to 15% more fuel cost than normal. Boilers need to be cleaned regularly on both the wet and heated sides in order to maintain desired thermal conductivity. The internal cleaning of the boiler should include a program of regular testing of the boiler water so that if an acid condition occurs it can be rectified. Also, unwanted chemicals in the boiler water will at times need to be removed. Usually, a cleaning concentrate is sprayed into boiler tubes where its high alkaline content neutralizes the acid sulphur of the hard crusts inside the fire tube walls.

Leaking steam traps are a source of considerable lost energy. For example, an orifice as small as ⅛ in. in a 5 psi steam system will permit the loss of about 4,000,000 btu for every 1000 hr of operation. All steam traps should be inspected periodically to assure they are working satisfactorily.

Exposed steam pipes should be well insulated. Periodic inspection should include the installation of new insulation in those areas where the insulation is missing or has been damaged.

It is sound practice to utilize thermostatically controlled valves on radiators. The thermostats should be checked annually. An improperly calibrated thermostat can be extremely costly (see Table 12.4).

It is a good idea to install an automatic thermostat control connected to an outside monitor in order to turn off the furnace when the outside temperature rises sufficiently.

Air filters need to be cleaned regularly or replaced. Outside air dampers seals should be checked so that cold outside air does not leak in when the dampers are closed. During the warm summer months the univent blowers should be turned off.

Hot Water Systems

Where domestic hot water systems are installed, the thermostat setting should be such so that the hot water delivered at the most remote point in the system is 105°F.

Table 12.4 Importance of Properly Calibrated Thermostats

Night temperature setting	Percentage fuel saved when temperature is lowered for 12 hr
68°F (20°C)	0
65°F (18.3°C)	4.5
60°F (15.6°C)	11.5
55°F (12.8°C)	18.0

In the hot water system, a time clock should be installed to control the water circulating pump so that circulation will be stopped during those times that the building is not in use such as nights and holidays.

Flow reducers installed on all hot water outlets including showers and basins will reduce the use of hot water at no inconvenience to the employees.

It is usually cost-effective to install a separate domestic water heater so that the boiler can be turned off when the heating season ends. When the boiler is turned back on, then of course the domestic water heater is turned off.

When procuring an electric water heater, an energy-efficient model should be specified. These have extra insulation resulting in some additional initial cost, but will result in significant operating cost savings. The water heater should be of the proper size for maximum energy savings. If the size is larger than required, then money will be wasted in heating and storing excess hot water. Of course, if an insufficient supply of hot water is produced, the installation will be a failure.

The water heater should be placed centrally and near to the points of service in order to minimize the heat loss in moving hot water from the heater to the point of service. For example, ¾-in. copper pipe installed in an environment of 50°F with 130°F water flowing through it will lose 30 Btu per foot of pipe per hour.

It is a good idea to insulate water pipes—hot and cold—to reduce heat loss and conserve energy. Pipe insulation comes in a number of forms, including wraparound flexible tubing, rigid foam, and rubber.

Even though an electric water heater has built-in insulation, heat will escape through the tank walls. Therefore it is a good idea to wrap insulation around the outside of the heater, reducing the amount of energy needed to maintain water temperature at the desired level (see Fig. 12.3).

A leaking hot water faucet wastes water, energy, and money, and should be repaired when discovered. Typically, a faucet that is leaking at the rate of 120 drops per minute will result in 429 gallons of hot water per month going down the drain. This would require 107 kWh per month, and at $0.08 per kWh would be a loss of $102.72 per year.

INSULATION

No matter how a plant or business is heated and cooled, it is not unusual to reduce the load on heating and cooling equipment by as much as 20–30% by proper insulation.

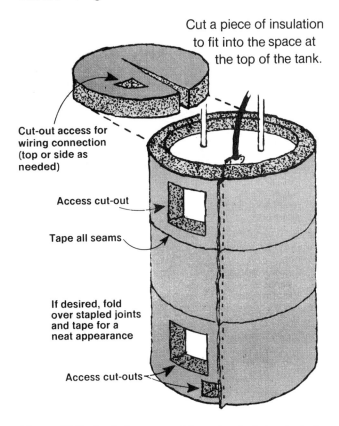

Cut a piece of insulation
to fit into the space at
the top of the tank.

Cut-out access for
wiring connection
(top or side as
needed)

Access cut-out

Tape all seams

If desired, fold
over stapled joints
and tape for a
neat appearance

Access cut-outs

Figure 12.3   Insulation prepared for a circular hot water tank.

The insulation installed should be based upon R values (insulation efficiency ratings; "R" stands for resistance to winter heat loss or summer heat gain). The higher the R number, the more effective the insulating capability. Figure 12.4 illustrates the recommended R values for attic floors, exterior walls, and ceilings over unheated space for five heating zones characterized by segments of the United States. Table 12.5 provides R values for various thicknesses of batts and loose fill insulation.

The most prevalent insulating materials available for use today include mineral wool, cellulose fiber, vermiculite and pearlite, and plastic forms (polyurethane, polystyrene, or urea formaldehyde). Of these, mineral wool, either glass or rock wool, is the most widely used type. It is available in these forms:

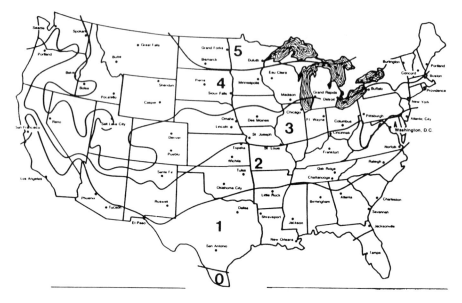

Recommended R Values

Heating Zone	Attic Floors	Exterior Walls	Ceilings Over Unheated Crawl Space or Basement
1	R-26	R Value of full wall	R-11
2	R-26	insulation, which is	R-13
3	R-30	3½" thick, will depend	R-19
4	R-33	on material used.	R-22
5	R-38	Range is R-11 to R-13.	R-22

Figure 12.4 Heating zone map. (Courtesy of U.S. Department of Energy.)

1. *Blankets.*   Rolls of insulation, with vapor barriers or without (unfaced).
2. *Batts.*   Similar to blankets but cut into 4-ft or 8-ft lengths.
3. *Pouring wool.*   Loose insulation that is poured into space between joists and studs.
4. *Blowing wool.*   Loose insulation that is blown into place using pneumatic equipment.

The following checklist can be helpful in connection with regular maintenance energy conservation audits:

Table 12.5 Inches of Insulation Required for Various R Values

| | Batts or blankets | | Loose fill (poured in) | | |
| | | | | | Cellulosic |
R value	Glass fiber	Rock wool	Glass fiber	Rock wool	fiber
11	3.5–4	3	5	4	3
13	4	4.5	6	4.5	3.5
19	6–6.5	5.25	8–9	6–7	5
22	6.5	6	10	7–8	6
26	8	8.5	12	9	7–7.5
30	9.5–10.5	9	13–14	10–11	8
33	11	10	15	11–12	9
38	12–13	10.5	17–18	13–14	10.5–11

Source: U.S. Department of Energy.

1. Consider adding (or installing) insulation above the upper floor ceilings in accordance with recommended R values. Note when piping is located in the heated and insulated envelope of the building, the problem of freezing pipes is eliminated.
2. Install insulation to all electric domestic water heaters.
3. Be sure that all steam piping is adequately insulated.
4. Consider providing additional exterior insulation on flat or low-sloped roofs.
5. When remodeling involves any exterior wall, consider installing insulation. This should include a vapor barrier installed toward the heated side of the insulation.
6. Consider installing roof vent fans which can both vent to the outside during the summer months and recirculate warm ceiling air down in the winter months.
7. Check ceiling vents to assure that warm air during the heating season is not permitted to escape.

Doors, Windows, and Skylights

Doors, windows, and skylights are sources of significant heat loss. This is because (1) glass is a highly heat-conductive material, (2) doors and windows that open necessarily have clearance all around them, (3) air will pass through the joints around window and door frames if they are not sealed tightly.

Storm windows and storm doors will cut heat loss at these points by about 50%. Insulating glass (two panes of glass sealed together at the edges) has approximately the same effect. Triple glazing (insulating glass plus a storm window) does of course provide even more insulation and is usually cost-effective in cold climates.

The maintenance analyst, in conjunction with the inspection of all window closures, needs to understand the application of both the U factor and the shading coefficient (S).

The U factor is the overall heat transmission coefficient expressed in BTU/(hr/ft^2/°F). It is the reciprocal of the sum of all resistances in an assembly, including air films.

The larger the U factor, the greater the heat transfer through the glass. For example, ¼-in. clear single-plate exterior glass has a winter U factor of 1.13 while ¼-in. double-plate exterior glass with ³⁄₁₆-in. air space has a winter U factor of only 0.69, thus permitting the transmittance of only 0.69/1.13, or 61%, as much heat as the single-plate glass.

An example illustrating the use of the U factor in computing the overall heat transmittance of the wall of a building might be helpful to the reader.

The wall to be calculated is 30 ft by 10 ft. It has two windows 5 ft wide and 4 ft high, and one door 3 ft wide and 7 ft high. Transmittance values are 0.08 for the opaque portion, 1.13 for the windows, and 0.43 for the two inch solid wood door. What is the U factor (total thermal transmittance) of the wall?

U_t = the total average thermal transmittance of the total wall area

A_t = the total exposed wall area of the building that faces heated space

U_o = thermal resistance of the opaque wall area

A_o = opaque wall area

U_w = thermal resistance of the window sash area

A_w = window sash area

U_d = thermal resistance of the door area

A_d = the door area

$$U_t = \frac{U_o A_o + 2U_w A_w + U_d A_d}{A_t}$$

Areas:

$$A_t = 30 \times 10 = 300 \text{ ft}^2$$
$$A_w = 2 \times 5 \times 4 = 40 \text{ ft}^2$$

$A_d = 3 \times 7 = 21 \text{ ft}^2$

$A_o = 300 - 40 - 21 = 239 \text{ ft}^2$

$U_t = \dfrac{(0.08)(239) + (1.13)(40) + (0.43)(21)}{300} = 0.245$

The shading coefficient is a numerical measure of solar heat gain on an area incorporating windows. The smaller the shading coefficient, the less the solar heat penetration. For example, a window facing the east with no shading would have a solar heat gain value of 50. However, with inside shading and overhang of 3 ft, the S value would be reduced to 20.

Window sash and doors should be weather-stripped on all sides and top and bottom. Several different types are available—spring bronze, flexible vinyl, foam rubber, felt strip, and others. Savings from weather stripping can be substantial.

All cracks around windows, skylights, and doorframes should be caulked properly. Deep cracks should be stuffed with mineral wool before the caulking is applied.

All doors, windows, and skylights should be inspected at least annually to assure that they fit tightly, that cracks have been sealed, that the glass is clean, and that cracked or broken panes are replaced. The following checklist should be used by the maintenance inspector when making his audit:

1. Skylights can be large energy wasters. An additional glazing of plexiglass is recommended.
2. Adjust outside door closers so doors cannot be propped open. Also adjust door closers so that doors will close faster in cold weather.
3. Doors and windows that are badly worn and cannot be weather-stripped or sealed effectively should be replaced.
4. Those entrances and windows exposed to prevailing winds may be protected by a windbreak. Consider plantings that will not only provide this protection but will enhance the aesthetic appearance of the building.
5. Windows that have no use for entrance of light or fresh air should be considered for closing in order to reduce the square footage of the glass area. An exterior panel over the window or bricking-in the window area would be cost-effective.

ROOFS

Roofing systems are important since usually they represent an industrial plant's largest exterior surface. Frequently, the roof can account for three times the surface area of the walls. The two criteria that have an impact on the energy costs of commercial roofing systems are the R value as already discussed and the surface color of the membrane. Black roofing membranes absorb heat, much of which is conducted to the interior of the building. On the other hand, white membranes will reflect up to 78% of the sun's radiant energy, thus requiring considerably less energy to keep the building cool.

The maintenance analyst, in addition to considering higher R value insulation, should consider lighter colors when replacing roofing or building a new structure where the inside of the building is maintained at temperatures between 68 and 78°F (see Fig. 12.5).

OTHER ENERGY CONSERVATION STEPS

Here are nine further considerations for conserving energy:

    1. *Energy savings controls.* Conventional full-voltage starting of large electric motors can require up to six times more current than is needed

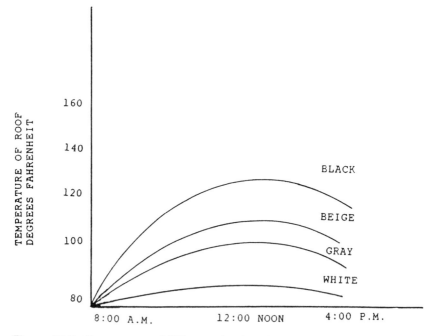

Figure 12.5 Temperatures of different colored roofs during the period 8:00 A.M. to 4:00 P.M.

for normal operation. Solid-state, soft-start energy savings controls may be installed on the load side of standard electro-mechanical motor controls. Electric power is conserved by sensing the motor slip. This signal is sensed and compared with a reference resulting in a signal to the logic to adjust the voltage. Energy savings of 10 to 40% from 30% full load to start-up is typical.

2. *Retrofitting old motors.*   Retrofitting older motors can result in significant savings. Energy-efficient electric motors used to replace old less efficient ones are usually cost-effective today in view of the rising price of energy. New motors manufactured today operate cooler and have much longer insulation life and bearing lubricant life. Most new motors will have improved characteristics in the stator and rotor cores and more copper in the windings. Those motors selected for replacement should be those on older equipment where the usage is high.

3. *Time clocks and thermostats.*   The use of time clocks on major load facilities including air handlers in order to reduce peak demand electric charges can prove fruitful.

   Thermostats should be set or programmed to 55°F in many of the plant areas just prior to the shift end in order to save energy during periods when the plant is not in use.

   Significant energy can often be saved by raising the chilled water temperature in the air conditioning chiller, and reducing the hot water temperature in hot water heaters.

4. *Heat transfer.*   The transfer of heat from refrigerating and air conditioning systems to provide building water heating requirements is a method that frequently has application to reduce considerably the water heating costs and the power costs. For example, one plant currently is heating approximately 11,000 gallons of water daily to 100°F by using the heat available from the plant's refrigerator system. In addition to the savings from the reduced cost of water heating, the company is experiencing a reduced cost, as well as increased life and efficiency improvement, of the refrigeration system because the lower temperature of the returning refrigerant reduces the head pressure on the compressors.

5. *Metal reflectors.*   The installation of metal reflectors behind radiators to reflect more heat in a work area rather than unnecessarily heating the wall can be beneficial.

6. *Solar energy.*   Another alternative that should be considered to produce hot water is harnessing solar energy. The majority of industrial and business users produce hot water with hot water or steam boilers

using mixing valves at the point of use or heat exchangers. It is generally conceded that solar energy systems can produce hot water economically to temperatures of 160°F to 180°F. In addition to the savings of solar heating the water, a large portion of the solid deposits that cause maintenance problems in boiler tubes may be removed easily prior to entering the boiler.

7. *Cogeneration.* Many manufacturing industries as well as the utility industry may find cogeneration attractive. This is the simultaneous production of thermal and mechanical or electrical energy from the same fuel.

8. *Energy recovery.* Those businesses, industries, and institutions that have considerable combustible trash in connection with their operation should consider the installation of a waste heat recovery system. Instead of burning high-cost fuels (gas, oil, coal) to generate steam, the trash that is being hauled away regularly (at a cost) can be burned to generate steam. A return of investment of approximately 30% has been reported on the installation of an incinerator system that effectively burns the waste leaving a clean, clear exhaust that meets federal and state regulations.

9. *Surge suppressors.* Most electronic equipment is designed to operate within a limited range of electrical voltages. Disturbances in the electrical supply to this equipment can adversely affect its performance.

   Voltage surges or spikes can originate both outside the plant on power lines or inside the plant. Utility switching can cause an external source of voltage surges, and a user within the plant cycling power on and off can cause an internal voltage surge.

   To prevent damage to electronic equipment, protection should be installed on the main electrical panel and on all electrical outlets serving electronic loads. Thus, lightning arrestors should be installed at the main panel and surge suppressors at individual outlets. Both of these devices direct the excess energy in a voltage surge away from the electronic equipment, usually to a ground, thus preventing damage to the equipment.

SUMMARY

Utilities management, embracing energy conservation, is a key responsibility of the maintenance function. There are two broad areas that represent opportunities for study in order to save energy:

1. The envelope where the business, industry, or institution is housed.
2. The process or operations that take place in connection with running the business, industry, or institution.

From the standpoint of the envelope, a major energy saver is insulation. Insulation of a proper type and thickness needs to be applied between joists of the ceiling of the top floor and between studding of all exterior walls of the building. In addition, heat loss should be retarded by insulating such facilities as domestic hot water heaters and service lines such as steam pipes, hot water, and so on.

Loss of heat through windows, doors, and skylights needs to be minimized by providing insulating glass, weather stripping, and caulking.

Another principal consideration in studying the envelope is the avoidance of excess use of electricity, oil, and gas in the lighting, heating, and air conditioning of the building. Periodic review of the lighting requirements throughout the enterprise is always advisable. Modern lamp installations coupled with an energy management program utilizing programmable control not only for lighting but for heating, ventilating, and air conditioning will usually prove to be fruitful.

A third important area belonging to the building envelope that should be monitored regularly is the boiler/heating system. Boilers need to be cleaned regularly in order to maintain efficiency. Steam traps need to be inspected periodically to insure that they are not leaking and are working satisfactorily. Air filters need to be replaced at regular intervals.

From the standpoint of the operations performed throughout the plant, the principal area for study is the conservation of electric power used by the rotating machinery. Older equipment will inevitably be utilizing inefficient motors. If this equipment is scheduled for high usage, consideration should be given to retrofitting with modern energy-efficient electric motors.

On large (over 30 hp) motors, consideration should be given to installing power factor correction capacitors. This can result in considerable start-up power savings.

SELECTED BIBLIOGRAPHY

Capehart, B. L., and Capehart, L. C. "Efficiency in Industrial Cogeneration: The Regulatory Role." *Public Utilities Fortnightly,* Volume 125 No. 6 (1990).
Chandler, W. U., Geller, H. S., and Ledbetter, M. R. *Energy Efficiency: A New Agenda.* Washington, D.C.: American Council for an Efficient Economy, 1988.

Energyworks, Inc. *Energy-Efficient Products and Systems: A Comparative Catalog for Architects and Engineers*. New York: Wiley, 1983.

Gandhi, N. "Industrial Energy Management Using I.E. Techniques." Proceedings of the Spring 1989 Industrial Engineering Institute Conference, Toronto, 1989.

Harrell, Joe J. *Solar Heating and Cooling of Buildings*. Florence, KY: Van Nostrand Reinhold, 1983.

13

Plant Rearrangement, Minor Construction, and Subcontracted Services

In many plant rearrangements, a professional architect is employed to design a plan to achieve desired objectives. When this is the case, it is highly advantageous to have the plant engineer and the maintenance manager work with the architect to help assure that the best possible arrangement is realized. The more the regular employees are involved in the rearrangement and construction, the more likely there will be complete satisfaction. It is the employees who have an in-depth knowledge of both the electrical and mechanical systems that will be altered. Also, these are the employees who will be responsible for maintenance of the plant and facilities after the work is completed. When they participate in both the planning and the renovation and construction phases, favorable results will inevitably occur.

The architect and his support from industrial engineering are interested mainly in efficiency of production operation and frequently forget the importance of maintenance. As the plans for the plant rearrangement are being developed, the plant engineer and the maintenance manager need to think constantly in terms of maintenance. Thought needs to be given to storage space, sizes of piping, filters, power lines, and so on, floor space around a production facility adequate for preventive maintenance and repair, and many other factors.

At the conclusion of the plant rearrangement and construction, it is the maintenance manager who has the responsibility for the maintenance of all the production equipment and the building envelope.

THE PLANT LAYOUT

The layout of a plant is a dynamic situation, always subject to change. When a plant grows or its volume of demand diminishes or its product output changes in style or model, there is invariably a need to rearrange the existing layout. The principal objective of effective plant layout is to develop a production system that permits the manufacture of the desired number of products of the desired quality at the least cost and in time to meet the purchaser's requirements.

Although it is difficult and costly to make changes in arrangements that already exist, management should require periodic review of every portion of the existing layout. The recommended rearrangement of the layout is usually a function of the industrial engineering department and the implementation of the proposed layout is usually handled by maintenance. Most modern managers recognize that poor plant layouts result in major costs. Unfortunately, most of these costs are hidden and consequently cannot readily be exposed. The indirect labor expense of long moves, backtracking, delays, and work stoppages due to bottlenecks are characteristic of a plant with an antiquated layout.

Layout Types

In general, all plant layouts represent one or a combination of two basic layouts. These are product, or straight-line, layouts and process, or functional, layouts. In the straight-line layout, equipment is located so that the distance from one operation to the next is minimized for any product class. Thus, in this type of layout, it would not be unusual to have a turret lathe located next to a thread grinder. Parts coming off the turret lathe would be in position for the next operation, grinding threads. The reader can appreciate that rearrangement under product grouping results in more difficult procedures for the maintenance department since there can be (and usually is) significant differences in the utility, foundation, and space requirements of adjacent equipment. Thus while product grouping usually is more efficient, it usually is also more costly to install.

Process layout is the grouping of similar facilities. Thus, all turret lathes would be grouped in one section, department, or building. Milling machines, drill presses, and punch presses would also be grouped in their respective sections. Under a process layout, the rearrangement of facilities is simplified for the maintenance department since economies of scale will exist as like facilities are grouped together in an orderly arrangement.

The maintenance department usually is given a layout of the proposed rearrangement developed by industrial engineering. The layout frequently is drawn, either manually or by CAD, to a scale of ¼ in. = 1 ft. Although the layout will show exactly where a facility is to be located in relation to building columns, walls, and windows, usually it does not provide detailed information as to utility requirement locations for power, gas, oil, air, water, and so on. The maintenance planner should sketch these requirements on the layout provided as well as any special foundation requirements that will be needed.

THE ELECTRICAL LAYOUT

The maintenance department will need to consult plant engineering in connection with details of the electrical layout as shown on working drawings. Through the use of symbols and lines, there should be no question of what is required.

The specifications of the electrical system represent the written description of what will be required of the maintenance workers or outside contractors. Together with the drawings, they provide the details of the work order for the installation of all utilities.

The following list of electrical symbols are representative of what is being used today:

————	conduit concealed in wall or ceiling
— — —	conduit concealed in floor
– – – – –	conduit exposed
—×—	flexible metallic armored cable
—//—	two-wire circuit
—///—	three wires
—////—	four wires
—T—	telephone conduit
—TV—	television antenna conduit
—S //—	sound system conduit; the number of cross marks indicates the number of pairs of conductors
Ⓕ	fan coil-unit connection
Ⓞ	motor connection

M.H.	mounting height
[F]	fire alarm striking station
(G)	fire alarm gong
(D)	fire detector
(SD)	smoke detector
(B)	program bell
(C)	clock
(M)	microphone, wall mounted
[M]	microphone, floor mounted
[S]▷	speaker, wall mounted
(S)	speaker, recessed
▲	telephone outlet, wall
(T)	television outlet
▭	fluorescent fixture
▯ C	incandescent fixture, recessed
◯ B	incandescent fixture, surface or pendant
⊢◯ A	incandescent fixture, wall mounted (on fixture symbols, the letter outside the symbol denotes switch control)
/A\	indicates fixture type
⊖	receptacle, duplex-grounded
⊖ WP	receptacle, weather proof
⊖ S	combination switch and receptacle
(·)	receptacle, floor type
S	switch, single pole
S_3	switch, three way
S_4	switch, four way
S_T	switch, toggle with thermal overload protection
S_P	switch and pilot light
[·]	push button
▱	buzzer
(J)	junction box
▭⌐	disconnect switch
⊠	starter

The National Electrical Code

It is important that the electricians working in the maintenance department be familiar with and observe the mandatory rules of the National Electrical Code. This code is published by the National Fire Protection Association (NFPA). Observance of the code's rules and regulations will help prevent fire and explosion hazards resulting from improper installation of electrical wiring, selection of materials and parts, and quality of work.

This code provides both mandatory and advisory rules in connection with grounded conductors, branch circuits, feeders, over-current protection, wiring methods and materials, and equipment for both special and general use.

INSTALLATION OF COMPRESSORS AND PUMPS

It is quite likely that plant rearrangement will at times require the relocation of compressors and pumps that supply compressed air and liquids under pressure to a facility.

For the relocation of this equipment, both compressors and pumps should be installed in a clean, well-lit room with adequate space around each piece of equipment so that it can be inspected easily, cleaned, and provided planned maintenance. Planned maintenance implies that there needs to be ample room for removing pistons, rods, and intercooler nests.

The intake of compressors should not be in close proximity to exhaust pipes discharging dust, steam, or other waste which may be brought into the compressor by intake suction.

A compressed air system that is designed improperly can result in a decrease in the efficiency of air-driven tools and facilities. Also, there can be a continuing increase in power cost resulting from an air compressor operating at a higher discharge pressure to compensate for pressure drops.

The added cost associated with compressing an increased air capacity to compensate for high pressure drops and leaks can be estimated by multiplying the horsepower of the motor, a constant (0.746), the operating hours per year, the cost of electricity per kilowatthour, and the decimal equivalent of the additional air capacity. For example, a 90% efficient 60-hp compressor that operates 5000 hr/yr (where electrical costs are $0.055 per kilowatthour) and compresses an additional 15% of air to compensate for high pressure drops and leaks would develop an added cost of

$$\frac{60}{0.9} \times 0.746 \times 5000 \times 0.55 \times 0.15 = \$2052$$

An approximation rule is that increasing the compressor discharge pressure by 2% results in an increase of 1% in horsepower required to operate the compressor.

The best way to reduce pressure drop without interfering with machine performance or increasing compressor maintenance and higher operating costs is through the proper selection of compressed air accessories and sizing of the compressed air piping network.

In the selection of an air compressor, care must be exercised to procure one in which the maximum rated discharge is sufficient to operate the air machinery installed while overcoming the inherent pressure drop in the piping and filtering systems. The maximum rated discharge pressure is that maximum pressure realized after all internal pressure losses have been accounted for. Usually, the installation of the air compressor will include an air dryer and oil removal filters. It is important that safe moisture level limits be controlled to protect both the compressed air system and those processes utilizing the compressed air. In a compressed air line, moisture becomes fluidized to a mist by turbulent air flow. The propelled droplets impact at piping elbows, valve discs, and on air motor blades, causing corrosion pits resulting in a short service life. As the filters trap contaminants, the pressure drop will increase. To assure that the filtering system operates within the expected pressure drop range, it is necessary to replace the filters periodically.

The piping system should be planned so that the pressure drop between the compressor system and the point of air consumption is low. Of course, the system should not contain leaks. Some surveys point out that it is not uncommon to lose as much as 10% of the total compressed air produced because of leakage. The maintenance effort should inspect piping connections, air hoses, and connecting fittings regularly for leaks. Repairs should take place promptly since air leaks, as pointed out, are costly.

It is a good idea to dimension the piping system so that future expansion of air utilization will not result in significant pressure drops. Small transmission lines will result in pressure losses because of higher friction, especially when the pipes are long. The engineering maintenance manager should keep in mind the basic Fanning equation:

$$h = \frac{flv^2}{2\,gD}$$

where

h = loss of head in feet of gas at average pressure and temperature in the pipe

f = friction factor

l = length of pipe, ft

v = average velocity of gas in pipe, ft/s

g = acceleration due to gravity, 32.174 ft/s^2

D = internal diameter of pipe, ft

It is apparent as l increases and D decreases, the loss of head increases. This will cause inefficient operation of power-driven tools.

In the installation of pipelines, avoid elbows and sharp bends as they result in added friction and restriction of flow. Provide plenty of standardized outlets with a shut-off valve at each outlet in the work area to avoid the use of extra long hose. The pipeline should be installed so as to drain to traps that have been placed at regular intervals for the removal of moisture.

In most cases, the best installation of the piping system is in the form of a closed loop circling the plant area where the air consumption will take place. This design allows a uniform supply of compressed air at usage areas, where air will be conducted to branch lines from two directions. This main line will be sized to accommodate the total airflow required. From the main line, branch lines will be run to specific areas of air consumption. Then feeder lines will be attached from branch lines to individual points of use.

The intake duct of the compressor should be installed on the exterior of the coolest side of the building (usually the north side) at an elevation that is free from the collection of debris and dirt. The intake duct should not be located near an area that would result in moisture or dust being drawn into the duct. The further away the intake duct is from the compressor, the larger should be the duct diameter. For every 10 ft away from the compressor, the pipe should be 1 in. greater in diameter than it is at the compressor.

The air-receiving tank should be located close to the compressor in order to keep the discharge pipe short. In the event the compressor is connected to a central air main to which other compressors are connected, a safety valve should be installed between the compressor and the first valve in the pipeline in order to protect the compressor in the event the first valve was left closed when the compressor was started up. Drain cocks should be installed on all air receivers near the bottom to permit regular periodic draining.

It usually is advantageous to install an air aftercooler near the compressor but ahead of the air receiver. The air aftercooler will condense the hot moist air from the compressor before it gets into the pipeline. Moist air in the pipeline will cause water hammer resulting in leaking joints and disrupting performance of pneumatic tools.

In steam-driven compressors, steam piping should drain toward the compressor so that water does not accumulate in low spots or pockets in the pipes and be carried into the cylinders. In order to furnish drier steam, thus providing a safer and more economical installation, the use of a steam separator and trap is recommended. A stop-valve should be installed at the point where the steam pipe branches from the main line. This can be used to shut down the compressor to make repairs such as repacking or regrinding the throttle valve.

Also, a drain should be installed in the steam pipe near the compressor. All steam piping should be well-insulated to prevent heat loss.

Belt driven compressors should use a belt ½ in. to 1 in. narrower than the face of the belt wheel. The belt should be tight enough to prevent slippage, but should not be too tight as this would place an unduly heavy load upon the bearings causing them to heat and subsequently freeze.

FOUNDATIONS

Concrete foundations of one part cement, two parts sand, and five parts rock aggregate are quite satisfactory. In order to provide added strength to the foundation when installing large compressors, pumps, or equipment, steel bars embedded both lengthwise and crosswise are helpful.

The foundation bolts can be located properly through the use of a wooden template. 3-in. pipe placed around these bolts will allow adequate distance for moving the bolts slightly when the machine is installed. A temporary dam made of boards or damp clay can be built around the edge of the base. An adequate quantity of grout, made of one part Portland cement and two parts sand, should be poured into the basin built around the base of the facility being mounted. The grout will flow under the base of the machine, filling the space left around the anchor bolts and between the base of the machine and the surface of the foundation. Thus, the machine frame will have equal bearing at all points on the foundation after the setting of the cement.

After the grout is set, the dam may be removed, the grout smoothed off around the machine base, and the anchor nuts screwed down tightly. The machine should not be run until the foundation has set completely.

MAINTENANCE SUBCONTRACTED SERVICES

The alert maintenance manager should consider the possible advantages of contracting certain maintenance work. Often, it is cost-effective to contract for such work as janitorial services, roof repair and replacement, painting, building construction, and repair of specialized equipment, such as computers, word processors, and sophisticated plant production facilities.

The purchased labor contracts for work should carefully spell out the quality of supplies, materials, and replacement parts to be used. Also, they should define the time period in which the work must be completed. The principal disadvantage of subcontracting maintenance is that the purchaser lacks control over the work. Thus, both quality and timing can be adversely affected.

Contractor Limitations

Before the maintenance manager decides to use a contractor to handle a certain maintenance function, possible limitations must be considered. This includes the availability of reliable contractors and the cost to perform the work.

Often, a qualified contractor is not available in the geographical area and, if one is, their schedule may be such that they are unable to take on the volume of work desired. Of course, the maintenance manager must have assurance that the contractor is capable of performing quality work in the area that is affected. If there is no available qualified contractor in the area, the maintenance manager may consider importing maintenance talent. This procedure can result in acquiring highly competent personnel, but the cost often is prohibitive and the analyst will usually find it cost-effective to develop and train the in-plant crew to perform the work.

Today, the competitive nature of all aspects of industry, the liberalizing of unions' restriction on the diversification of duties and responsibilities of its members, and the increasing importance for all employees to provide a "fair day's work" often require the expansion of a typical maintenance employee's job to include more than one craft. For example, today machine operators in many plants are taking on a significant role in connection with preventive maintenance of their work stations. In a similar vein, the plant carpenters may be used to perform window-glazing maintenance, and the plant electrician at times may be used with a crew of pipe fitters.

Of course, the amount of work required by a specified date is a major consideration in deciding whether the work should be done in-plant or on a contractual basis. An estimate of the potential available hours of the plant's existing maintenance force is the first information needed. If this is equal to or greater than the time required to perform the needed work, then the work should probably be handled in-house. On the other hand, if the maintenance work is of a specialized nature requiring unique equipment and skills and the volume is insufficient to keep an employee fully occupied for a period of two to three days, usually it would be cost-effective to have the work done on a contractual basis. For example, to grind the valves of a special machine requiring a specialized grinding facility normally would be subcontracted.

A major reason for subcontracting is to handle those assignments where insufficient labor exists to complete the work by the time required. Since the budgeting of maintenance personnel typically is based upon average loads (as opposed to peak loads), there will be times when contractors should be used to bring back logs of work to levels commensurate with normal plant maintenance work schedules.

Finally, because of efficiency of performance, contractors are often able to perform certain maintenance work at a considerable savings when compared to having the work done in-house. Work such as roofing, laying carpet or tile, laying block or brick, and customized machining is often performed in a fraction of the time required when done in-house.

Making a Cost Appraisal

Before making a decision to contract work, the maintenance supervisor should know what the cost would be to do the maintenance in-house. This includes:

1. Direct labor costs, including fringe benefits. These costs are determined by multiplying the appropriate labor rates (including fringe benefits) by the estimated hours to perform the work.
2. Materials. These costs include all spare parts and replacement materials consumed in doing the maintenance work.
3. Fixed and variable burden. Usually, the company established separate rates for the fixed and variable burden. Fixed burden includes those continuous costs that take place independent of the volume of work per unit time. Thus, fixed burden includes administrative salaries, building rent or mortgage payments, building depreciation, insurance, and so on. Variable burden fluctuates as the volume of work per unit time changes. Variable burden includes the use of power, indirect labor, specialized tooling, maintenance facilities, and others.

In maintenance work, often both fixed and variable burden are assigned on the basis of the maintenance direct labor cost. Materials costs include not only the cost of those materials purchased but also the cost of inventorying them and the cost of their acquisition. The direct labor times are best estimated using the universal indirect labor standards approach discussed in Chapter 4.

Working with Contractors

The day-to-day business of working with contractors is handled directly by the purchasing department, with interaction by engineering maintenance. Most maintenance contract work is routine and can be scheduled regularly. Typical work done in this manner is elevator and escalator maintenance, solid waste removal, pest control, and janitorial services. Often it is helpful for the maintenance manager to provide supervision of the contracted labor to assure that there is a smooth working relation between maintenance and production workers. This supervisory support can be effective in providing a good relationship between in-house maintenance and contracted maintenance workers.

These positive relationships tend to assure that the contracted maintenance is performed on schedule and at a good quality level.

Of course there can be considerable contracted maintenance that is not regularly scheduled. This work is covered by contracts occasionally issued to reduce maintenance backlogs or to perform work for which in-house maintenance does not have the adequate experience or facilities to satisfactorily carry out the work.

In unscheduled maintenance, it is desirable to know those organizations that have proven capability for the work to be performed so that only quality contractors are contacted for submission of quotations. Where repeated work is required, it usually is fruitful to place open-ended contracts for maintenance labor. These contracts are used primarily as a means to maintain backlogs of required work at some planned level.

Selecting Qualified Contractors

Usually, the maintenance manager will find several local sources that are anxious to perform the contracted maintenance. Certainly, the ideal contractor is one who does quality work in accordance with a predetermined cost and time schedule. It is helpful for the maintenance manager to maintain an active file of all those contractors who can be of service to the company. The file on each contractor should provide an estimated point value for the following criteria:

A. Quality of work
   1. Experience of workmen
   2. Special skills or capability including special equipment
   3. Reputation
B. Adherence to schedule
   1. Availability
   2. Reputation
C. Cost
   1. Base rates of workers
   2. Use of incentives
   3. Discounts on spare parts or materials

For example, A, B, and C can be evaluated on a scale from 1 to 10. A rating of ten for quality of work would indicate the contractor had a team of highly qualified and experienced workmen who had those special skills working with the most favorable facilities and performed the highest quality work. This concern will have a favorable reputation in the region for doing only the best quality work.

A 10 rating for adherence to schedule would be given to those contractors who demonstrate internal control of their backlogs to assure that all scheduled work is performed on time or earlier. They would have a reputation of providing good service in connection with the scheduling of work and reliability in meeting schedule dates.

A 10 rating for cost would indicate the firm has a record of almost always providing the lowest cost among competitive bidders. The company has base rates, including fringe benefits, which are average or below average in the areas evaluated. They may also have an incentive system in effect that compensates workers in proportion to output. They have good relations with their suppliers and are able to qualify for significant discounts, because of economies of scale, in the purchase of materials, components, hardware, supplies, and so on.

Once these records are established for potential contractors, it is relatively easy to determine which is most appropriate (usually the one with the highest total point value). It is important that these evaluation records be updated regularly to assure their continued validity.

TYPES OF CONTRACTS FOR PERFORMING MAINTENANCE WORK

There are several types of contracts that can be executed for the performance of maintenance work. The maintenance manager should be aware of these various alternatives and the advantages and disadvantages of each type. Most contracts are either

1. Fixed price, or lump sum, contracts
   a. Unit price contracts
2. Cost plus fixed fee contracts
3. Cost plus percentage fee contracts
4. Guaranteed maximum contracts
5. Purchased labor contracts
6. Time and materials contracts

Fixed price contracts, as the name implies, requires that for the work completed, the contractor receives a payment that has been mutually agreed upon by the customer and the supplier prior to the completion of the contract. The price is usually the result of a competitive quotation submitted in advance of beginning work on the job.

The principal advantage of this type of contract is that the cost of the maintenance is known before work is begun. The responsibility for completion of the contract rests solely with the contractor. This type of contract provides the opportunity to introduce changes, accompanied, if necessary, by an alteration in the contract price. Fixed price contracts are usually used if the work to be accomplished can be clearly defined and specified.

A variation of the fixed price contract is the unit price contract, wherein units of output can be measured easily but the project is not estimated at the time of letting. For example, a paving contract may be estimated initially at an exact price per cubic yard, but the total fee will not be estimated at the time of letting. Projects that often are based on unit price contracts include roofing, paving, earth removal, and so on.

In the case of the cost plus fixed fee contract, the contractor is compensated for the cost of all work performed and is paid an established fee for overseeing the work. This kind of contract is quite flexible since alterations in the work to be done can easily be accomplished. Thus, there can be a continual effort only to perform work that needs to be done, which in maintenance work is usually not completely known until after the job is begun. Of course, in this type of contract, one does not know in advance what the actual cost of the contract will be.

The cost plus percentage fee contract is similar to the cost plus fixed fee contract except, instead of a fixed fee added to the cost of doing the work, the contractor is paid a fee that is determined by a percentage of the cost of all work performed. This type of contract also facilitates alteration of the work to be done at any time, since the charge is not fixed. The actual cost of this type of contract is not known in advance.

The guaranteed maximum contract is similar to both the preceding cost-plus contracts. Here, the contractor is compensated for the cost of all work performed, however, he is guaranteed a maximum figure. Any savings below this top cost figure are shared between the contractor and the company. These savings are usually shared on a fifty-fifty basis. The advantage of this type of contract is that the maximum cost is predetermined. The contractor also has an incentive to perform better since any savings will be shared. The disadvantage is that the total time for completion may take longer since the contractor usually will want to complete his working drawings before developing a guaranteed maximum figure.

Under purchased labor contracts, the maintenance supervisor purchases specified personnel at predetermined hourly rates in the quantity needed to perform the desired work. Thus, the buyer has considerable flexibility since only that type of labor needs to be purchased. Frequently, this is skilled specialty labor. The principal disadvantage of this type of contract is that the purchaser has little control over the quality of personnel provided or their performance.

For time and materials contracts, the purchaser pays a predetermined rate for each hour of labor provided plus a charge for each item of material furnished. The labor rate, of course, includes a mark-up for the contractor, as does the charge for materials, to cover procurement and storage costs. The advantage of this type of contract is its flexibility. Only work that needs to be done and only the required materials are utilized. Of course, in this type of contract, the total cost for the maintenance is not known in advance.

The decision of which type of contract to execute is usually based on the prior experiences and relationships between the business and various contractors, the urgency of the work to be completed, the extent of the knowledge in advance of what needs to be done, the estimated cost, and the nature of the project.

It has been my experience that in both normal and depressed times, fixed price contracts usually will save the purchaser money. In robust times, time and materials contracts are usually economically advantageous.

In any significant maintenance or construction project that is contracted, the purchasing company, through the maintenance program, should provide a supervisor, who is intimately familiar with the work to be performed. The supervisor should have been involved in the design of the project (if a construction project) or work in the area in which the maintenance is being performed. This individual will be the principal contact for the contractor and will work regularly with the contractor performing the following duties:

1. Review with the contractor details and specifications of the project and provide explanatory details where needed.
2. With the contractor, introduce the outside personnel to the company's employees involved at the job site.
3. Provide regular inspections as the work progresses to assure quality of workmanship and materials.
4. Work with the engineer from the contractor's company in connection with changes and modifications related to the project and see that marked-up or new drawings are provided for quotation changes.
5. Prepare punch lists and make final acceptance of the project on behalf of the company's management.

INSURANCE, GUARANTEES, AND LIENS

Insurance

Contractors should carry ample insurance to indemnify and hold harmless the owner while engaged on a job in a plant. Suggested minimal insurance for contractors includes

1. Worker's compensation and occupational disease insurance
2. Employer's liability insurance ($100,000 per person)
3. Comprehensive general liability insurance and contractual liability insurance
   a. Bodily injury ($500,000 per person; $1,000,000 per occurrence)
   b. Property damage ($1,000,000 per occurrence)

Guarantees

Guarantees should be required to cover failures in any installation due to defective workmanship or materials. It is always advisable to obtain a written guarantee for protection to the owner.

Liens

A contractor should not receive final payment for any part of retained percentages until a waiver of lien for materials, labor, and any other changes associated with the job he has performed has been provided.

SELECTED BIBLIOGRAPHY

Blocher, James D., and Chand, Suresh. "Resource Planning for Aggregate Production." In Gavriel Salvendy, Ed., *Handbook of Industrial Engineering*. 2nd ed. New York: Wiley, 1992.

Piper, J. "Facility Assessment Survey." *Building Operating Management*, Vol. 35, No. 9 (1988).

Piper, J. Retrofit's First Step; Facility Assessment, *Building Operating Management*, Vol. 37, No. 2 (1990).

Rosaler, R. C., and Rice, J. O. *Standard Handbook of Plant Engineering*. New York: McGraw-Hill, 1983.

14

Fire Protection and Controlling Residual Waste and Contaminated Storm Water Discharge

FIRE PROTECTION

There are approximately 150,000 fires per year in industrial plants, storage facilities, and mercantile and office establishments. The maintenance manager may be charged with the responsibility of fire protection for the plant. Fire prevention is an important activity that must be carried out in every industrial activity. The maintenance department often carries the responsibility of this effort. This responsibility includes (1) regular inspection for fire hazards and (2) preparation for controlling any fires that take place.

Regular inspection is performed to reduce the fire hazard. Preparation for fire control involves not only having the means to extinguish fire, but also providing installations and operating procedures so that once a fire is discovered there are provisions for safe and easy egress by doors, stairwells, fire escapes, and other emergency exits.

To prevent fires, consider the environmental situations known to be the cause of fires. This includes electric wiring and electrical equipment, the handling and storage of flammable liquids, including gasoline, the use of equipment and facilities that generate high temperatures, such as welding, soldering, grinding, and so on, and the utilization of flammable materials, such as spray paint.

A fundamental rule in fire prevention is that good housekeeping must prevail if fires are to be avoided. The following rules need to be observed:

1. Tight fitting covers on metal containers should be utilized in the collection and temporary storage of all flammable waste. These containers should be emptied at least daily.
2. Flammable liquids should be kept in tanks or tightly closed drums outside the building. If these liquids are wastes, they should never be poured down drains; they should be stored until safely disposed of.
3. Containers for rags and oily waste should be made of metal and have self-closing lids.
4. Rubbish should not be allowed to accumulate in the plant; it should be moved regularly outside the plant for subsequent removal.
5. All areas adjacent to the outside of the building should be kept clear of combustible materials, such as rubbish and dry bushes and grass.
6. Those liquids having a low flash point, such as alcohol, gasoline, acitone, paint thinners, and so on, need to be stored in identified safe areas and issued only in safety cans having tight self-cleaning caps with fire screens.
7. Smoking should never be permitted in areas such as storerooms, stockrooms, and plant areas where flammable liquids and materials are used.

Organization Considerations

All supervisors must be informed regularly as to details of the buildings and grounds of the organization. They need to be cognizant of such facilities as main control valves that regulate water and fuel supplys, pumps, hose houses, hydrants, check valves, and so on. There should be an established plan in connection with the utilization of water from available supplies (both public and private), such as streams, wells, ponds, tanks, and water mains.

Everyone must have general knowledge of all existing fire extinguishing equipment, including its application, limitations, and maintenance as well as its locations throughout the plant. Also, all supervisors must be knowledgeable on the location, operation, and maintenance of alarm systems, automatic sprinkler systems, and any other system, such as fog or foam, that may be installed for fire protection.

Most importantly, there needs to be a well-known, established procedure to be followed when a fire breaks out. Periodic fire drills should be exercised to assure that all persons know their specific roles in times of such an emergency.

Classification of Fires

Fires are classified as class A, class B, and class C.

Class A Fires. These are fires fueled by paper, wood, textiles, and similar combustible materials. Class A fires can be extinguished using large quantities of water. Also, soda acid extinguishers, which contain bicarbonate of soda solution and sulfuric acid, are effective in extinguishing class A fires. A third extinguisher used for class A fires is the gas cartridge, containing water and carbon dioxide gas.

Class B fires. This class pertains to those fires involving flammable liquids, such as organic solvents, oils, gasoline, and grease. To extinguish this class of fire, it is necessary to provide a blanket over the fire to prevent its exposure to oxygen. Carbon dioxide extinguishers should be used for this class of fire. These extinguishers contain liquid carbon dioxide under pressure. The typical CO_2 extinguisher is operated by pulling a pin and opening a valve. The 15-lb extinguisher will discharge a stream of carbon dioxide 6 to 8 ft for a period of approximately 45 seconds. Class B fires can also be controlled using foam extinguishers, which contain a solution of aluminum sulfate and bicarbonate of soda. This extinguisher is operated by turning it end-for-end. The 2½ gallon size will discharge a stream 30 to 40 ft for a period of 50 to 55 seconds.

Class C fires. These are electrical fires. Here, a nonconducting extinguishing agent is necessary. Since water is a good conductor, streams of water should not be used to extinguish electrical fires. For class C fires, dry chemical extinguishers are recommended. This extinguisher contains bicarbonate of soda, dry chemicals, and a cartridge of carbon dioxide gas. This extinguisher is actuated by pulling a pin, opening a valve, and squeezing a trigger. The 30-lb size will discharge about 15 ft for a period of 20 to 25 seconds.

Fire Fighting Procedure

Upon discovery of any fire in the plant, the fire alarm should be sounded to bring the fire brigade to the specific area. Fire doors, as well as all other doors and windows leading to the area, should be closed. All blowers, ventilators, and conveyors should be turned off. Unless it is obvious that the fire can be extinguished quickly, the fire department should be notified.

Those who first discover the fire should take steps to contain and extinguish it, using the correct extinguishers based upon the type of fire. Remember that water is the best extinguisher of fires involving solid materials if the supply is adequate.

CONTROLLING RESIDUAL WASTE AND CONTAMINATED STORM WATER DISCHARGE

The maintenance function of many businesses and industries is now required to assume responsibility for the reduction, control, transportation, and removal (or the recycling) of residual waste, possibly including the control of contaminated storm water discharge. Residual waste may be defined as any nonhazardous industrial waste, including by-products, co-products, and abandoned products; expended materials, and contaminated soil, water, and residue. It does not include the following:

1. Materials directly recycled or reused by the manufacturer
2. Products, unless abandoned or disposed
3. Co-products to be sold or transferred for use in lieu of a manufactured product or to be used by the manufacturer in lieu of a product, unless abandoned or disposed
4. Processed waste that meets definition of co-product
5. Materials from the slaughter and preparation of animals and fish

The volume of residual waste generated by industry is highly diversied. Typical examples of residual waste include ash generated by coal-burning plants and residual waste incinerators, ceramics, gypsum board, linoleum, leather, rubber, textiles, glass, pesticides, fertilizers, pharmaceutical waste, detergents and cleaners, paper, photographic film, foundry sand, shavings, and waste from the manufacture of lime and cement.

Residual waste can be disposed of in only three ways: it can be burned, it can be buried, and it can be composted. In Pennsylvania, approximately 400 facilities had permits to process or dispose residual wastes in 1993. This included approximately 50 incinerators, 120 waste landfills, 170 agricultural utilization facilities, and 35 municipal waste landfills.

The effective operation of a pollution prevention plan is linked closely to the reduction and control of residual waste. Residual waste source reduction has always been a promising area on which to focus. This effort includes the reduction or elimination of the quantity or toxicity of residual waste generated. This may be achieved through changes within the production process, such as process modifications, material substitutions, improvement in materials purity, and packing and shipping modifications. Also, source reduction of residual waste may be achieved by increasing the efficiency of process machinery and recycling within a process. Thus sound industrial engineering management and housekeeping practice is helpful.

Shrinking landfill availability, landfill and incinerator siting difficulties, lack of markets for recycled materials, and states trying to close their borders to waste from other states provide motivation for legislation attempting to help relieve the waste problem. The National Waste Reduction, Recycling and Management Act, Title III was drafted with the goal of creating additional markets for recycled materials. It requires that all product packaging must

1. Be made of a material that is recycled at a 25% rate annually by 1985 (except paper which is covered separately), 35% in 1988 and 50% in 2001; or
2. Be made of at least 25% post-consumer material by 1995, 35% by 1998, and 50% by 2001; or
3. Be designed to be refilled or reused at least five times; or
4. Be reduced in weight or volume by 15% from the previous year.

The maintenance manager should endeavor to prevent waste from being generated. If it is not practical to prevent waste, then efforts need to be made toward reclamation. That portion of waste that cannot be economically reclaimed needs to be treated prior to disposal so that there will be both an environmental and economic benefit.

The principal benefits achieved through a waste source reduction effort include the following:

1. Waste reduction or elimination results in more efficient, cost-effective operation of all work centers throughout the plant.
2. Greater conservation of raw materials.
3. Disposal costs are reduced as waste is reduced.
4. A firm's liability associated with waste will decrease as less waste is handled.
5. The firm's image will be enhanced as its efforts toward environmental protection become apparent.

Regulation of Residual Waste

The risks posed by unpermitted and inadequate facilities demand comprehensive protective regulations. Most states have regulatory bodies authorized to enforce compliance with state laws and regulations that pertain to the handling of waste. For example, in Pennsylvania, the Pennsylvania Department of Environmental Resources Bureau of Waste Management enforces compliance through issuing permits to waste handlers, requiring them to submit reports and analyses, performing on-site inspections, investigating complaints, and prosecuting violators.

Firms are required to abide by an increasing number of rules and regulations established and controlled by departments of environmental resources and environmental quality boards. Although the following discussion is not complete, it is representative of what is in effect in most states today.

Any firm that generates more than 2200 lb of residual waste in any given month of the year must prepare a Source Reduction Strategy (SRS). This documentation provides a description of the source reduction activities proposed, a schedule for implementation, and the projected amount of the estimated reduction in waste generation.

Also, an SRS will be required to be submitted along with any application for a permit to treat, process or dispose of waste. SRS reports will be required to be updated any time there is a significant change in the type of waste being generated or in the manufacturing process that is producing the waste.

Source Reduction Program

Although the maintenance department is charged with the responsibility of the operation of a sound source reduction waste program, there must be a total commitment by management. Management should help assure that the source reduction program is a company-wide effort. This focus by management will be enhanced if the following steps are taken:

1. Make it evident that source reduction of residual waste is an important component of company policy. All employees should be encouraged to participate in the effort by submitting their ideas in writing to the maintenance department.
2. Those suggestions that prove to be successful should receive adequate visibility so that all in the company are aware of the idea and the initiator of the suggestion.
3. The maintenance department should establish realistic waste reduction goals based upon experience and analysis of existing data.
4. Initiate and maintain a good waste accounting and auditing system.
5. Develop the true cost of waste and associate these costs in the overhead of those work centers responsible. Then have methods engineering develop means to eliminate or reduce the waste generated.
6. Maintenance should prepare an annual report for management in which results of the waste source reduction projects of the past year are summarized. The maintenance manager should assure that the program is monitored and updated regularly.

CONTROLLING CONTAMINATED
STORM WATER DISCHARGE

Since 1992, the Environmental Protection Agency (EPA) has required facilities to apply for a storm water permit, intending to reduce or, ideally, prevent contaminated materials from mixing with storm water and subsequently flowing into streams. In the vast majority of industrial plants, the maintenance department is charged with the responsibility of establishing and operating this pollution prevention plan.

The maintenance analyst, in developing a sound storm water pollution prevention plan should organize and head a pollution prevention committee which will report to the chief plant engineer. Once the committee is formed, it should identify the point source and possible routes of all contaminants throughout the plant. This information is obtained by making a thorough and detailed inspection of the entire plant. This visual inspection will evaluate any leaking machinery, pipes, tanks, storage vessels, and containers. An evaluation will be made of the raw materials used to determine if any of these are a possible source of pollution. In those cases where a possibility exists, consideration should be given to using alternative materials having a lower concentration of pollutant. It is a good idea to record all obtained information on a site map of the plant as well as in notes for subsequent analysis. In many plants today, the opportunity exists to reduce the quantity and severity of pollutants inventoried and to control them better.

Once the pollution prevention committee has completed its inspection of the plant and has made a record of the location of all pollution sources and where they could be discharged, the company should incorporate the following management practices to minimize the chance of introducing pollutants into storm water discharge.

1. Maintain an active pollution prevention plan. This involves incorporating storm water pollution prevention with the company preventive maintenance plan. Thus, qualified personnel will regularly inspect, test, and maintain storm water management equipment and instrumentation. This includes storage tanks, bins, pressure vessels, pumps, valves, piping, fittings, processing equipment, bulk materials storage areas, and materials handling equipment. Special attention is to be given to corrosion, pitting, cracks, leaks, seepage, and storage. Attention should also be given to possible pollution from high winds.

2. See that EPA sampling and monitoring regulations are complied with. These regulations require that storm water samples be collected from

a storm that drops more than 0.1 in. of rain within 72 hr of a previously measured storm. The following equipment is to be utilized:

    a. Rain gage to measure the amount of rain falling on the company site

    b. Flow measuring instrumentation to monitor the magnitude of any contaminated materials mixing with storm water

    c. Computers to develop a data bank where data can be recorded and stored

    d. Automatic samplers that take samples of storm water at key points without requiring plant personnel

3. Educate employees regarding the goals of the storm water pollution prevention and of their responsibilities in connection with the program. Lectures supplemented with appropriate material should cover pollution control laws and regulations, spill response, sound materials management practice, good housekeeping, and operations designed to minimize storm water pollution.

SUMMARY

Protection against fires in the plant is an important maintenance function. Good housekeeping and regular inspection for unsafe conditions and unsafe acts are the principal functions of the maintenance effort in connection with fire prevention. The plant management should provide regular training to all line supervisors as to the procedure to be followed whenever a fire is discovered, including the location and use of fire fighting equipment for the three classes of fires and details of the building regarding exits, fire walls, hydrants, and so on.

The disposal of residual waste is another responsibility of the maintenance effort. Residual waste can be disposed of in only three ways: burned, buried, and composted. By reducing and controlling the residual waste of a plant, the organization is able to reduce or prevent pollution. Thus, residual waste is an ongoing problem in most plants that needs to be addressed continually by plant management.

Another pollution control responsibility assigned to the maintenance department is that of controlling contaminated storm water discharge. There must be an on-going effort to prevent contaminated materials from mixing with storm water and subsequently flowing into springs, wells, or streams. Periodic inspections by the maintenance department to identify source points of contaminants, where they might go, and how they might be controlled should be a regular function.

SELECTED BIBLIOGRAPHY

Cheremisinoff, Nicholas P., Cheremisinoff, Fred E., and Perna, Angelo J. *Industrial and Hazardous Wastes Impoundment*. Ann Arbor, MI: Ann Arbor Science Publishers, 1979.

Forester, William S., and Skinner, John H. *International Perspectives on Hazardous Waste Management*. New York: Academic Press, 1987.

Freeman, Harry M. *Incinerating Hazardous Wastes*. Lancaster, PA: Technomic Publishing, 1988.

Malhotra, H. L. *Design of Fire-Resisting Structures*. New York: Surrey University Press, 1982.

Patterson, James W. *Waste Water Treatment Processes*. Ann Arbor, MI: Ann Arbor Science Publishers, 1978.

Przetak, Louis. *Standard Details for Fire-Resistive Building Construction*. New York: McGraw-Hill, 1977.

Ramalho, R. S. *Introduction to Waste Water Treatment Processes*. New York: Academic Press, 1977.

Shields, T. J., and Silock, G. W. H. *Buildings and Fire*. New York: Wiley, 1987.

15

Training for Engineering
Maintenance Management Work

We have defined the maintenance function as embracing all those actions that are necessary for retaining equipment and facilities in, or restoring them to, a serviceable condition. The three broad factors that determine the overall success of engineered maintenance management are designing for maintainability, engineering and management support, and maintenance personnel.

In Chapter 11, we learned that designing for maintainability covers the physical aspects of the equipment—including accessibility for normal maintenance and repair, test equipment and tools required to monitor facilities, and skill levels required to perform maintenance as dictated by design, accessibility, and location of test points.

Engineering and management support refers to the maintenance organization, paths of authority related to the maintenance function, and the logistics of the entire maintenance management system.

Maintenance personnel includes the skill level of the maintenance technicians, plant engineers, and maintenance engineers; their experience and technical proficiency; and their attitudes and enthusiasm for their work. In this chapter we focus on those training techniques for maintenance work that will result in the most effective maintenance personnel.

PRE-EMPLOYMENT TESTING

Selection of maintenance workers through pre-employment testing is recommended. History has proven this screening will provide more quality-conscious employees, who learn faster, are more productive, and stay on the job longer. States such as North Carolina and Pennsylvania, through their job service programs, monitor an aptitude test which can be helpful in matching workers with jobs. This testing can help to improve the quality of the maintenance work force and reduce the high cost of labor turnover.

Research has shown that, on the average, job performance and aptitude are linearly related. As aptitude increases, job performance improves. Also, research has proven that ability tests are the most accurate method for predicting future performance in entry level maintenance jobs.

The test used by the Job Service Program of the Pennsylvania Office of Employment Security is known as the General Aptitude Test Battery (GATB). This test measures nine aptitudes. These are identified by a letter code:

G—*General Learning*: Ability to understand or to catch on to instructions and underlying principles; to reason and to exercise judgment; closely related to performing well in school.

V—*Verbal Aptitude*: Ability to understand meanings of words and to apply them effectively; to comprehend language; to understand relationships between words; to understand the meaning of whole sentences and paragraphs.

N—*Numerical Aptitude*: Ability to perform arithmetic operations quickly and accurately.

S—*Spacial Aptitude*: Ability to think visually of geometric forms and to comprehend the two-dimensional representation of three-dimensional objects; to recognize the relationships which result from movement of objects in space.

P—*Form Perception*: Ability to perceive pertinent detail in objects or pictorial/graphic materials; to make visual comparisons and discriminations; to perceive the slight differences in shapes and shadings of figures, widths and lengths of lines.

Q—*Clerical Perception*: Ability to perceive pertinent detail in verbal or tabular material; to observe differences in copy; to proofread words and numbers; to avoid perceptual errors in arithmetic computation; also, a measure of speed of perception as required in many industrial jobs, even when said job is without verbal or numerical content.

K—*Motor Coordination*: Ability to rapidly and accurately coordinate eyes with hands or fingers while making precise movements with speed; to respond accurately and swiftly.

F—*Finger Dexterity*: Ability to move the fingers and to rapidly and accurately manipulate small objects with the fingers.

M—*Manual Dexterity*: Ability to move the hands easily and skillfully; to work with the hands in placing and turning motions.

Most maintenance work requires good scores in general learning, numerical aptitude, spacial aptitude, form perception, motor coordination, and manual dexterity.

IN-PLANT TRAINING FOR MAINTENANCE TECHNICIANS AND TECHNOLOGISTS

Maintenance is dynamic. New deficiencies in equipment are continually arising while old problems are in the process of being corrected; still other jobs are being completed. In order to cope with the tremendous number of variations that are characteristic of modern maintenance, it is wise to provide in-plant training so that maintenance technicians will be able to perform quality service and repair work in the least time. Those men and women selected for in-plant training should have the following attributes:

1. A high school or trade school education or its equivalent.
2. Training in at least one craft. This would be in that craft in which the person aspires to specialize, e.g., electronics technician, electrician, plumber, pipe fitter, carpenter, machinist, welder, etc.
3. A desire to improve oneself.
4. An ability to work with and get along with people.
5. An aptitude and interest in mechanical manipulation.

There should be a formal selection procedure that not only appraises the above basic qualifications, but tests the applicant's aptitudes in order to determine if the applicant is qualified to receive the training. In addition, the selection procedure should include an interview to determine the applicant's goals, attitudes, and work history.

The in-plant continuing education program for maintenance technicians should be under the direction of the plant engineer who holds the responsibility for the overall maintenance function. It is important that this person keep abreast of the skill requirements for the maintenance of all the equipment in the plant. These requirements keep changing with the passing of time. There is a tendency today to require more maintenance personnel that have specialized skills, such as electronics, fluidics, pneumatics, etc., and fewer maintenance generalists in skill areas such as plumbing, millwrighting and carpentry. Thus, the plant engineer should project the anticipated requirements of maintenance technicians at least three years into the future. These projections will be based on trends in the past and forecasts of the company as to anticipated production and the equipment needed to maintain the expected production requirements.

With knowledge of the estimated numbers of new people to be added with particular skills, the plant engineer can plan for those training programs which are needed.

When there is a change in the skills needed to fulfill a company's maintenance technician requirements, those personnel whose skills will be phased out soon should be given a training program in the area where added personnel is needed. This reeducation effort helps assure the security of hourly workers and does much to provide good labor–management relations. In addition these people already understand the goals, objectives, and aims of the maintenance function in the plant and therefore require a shorter training program.

If new people need to be employed, then those who have completed an apprentice training program or equivalent in the skill areas needed should be considered. Some high schools and post–high school technical institutes provide excellent programs that give a well-balanced curriculum that specializes in electronics, electrical, and hydraulics maintenance. These programs provide not only ample training in the particular skill area, but also necessary related instruction in such areas as mathematics (algebra, geometry, trigonometry, statistics), computer programming, drafting and computer-aided design, engineering graphics reading, and effective speech skills.

Inadequate training can make many maintenance operations dangerous. For example, many maintenance workers are obliged to perform welding duties even though they generally are less well-trained and well-protected than production workers. They are also much more likely to work in tight or confined spaces and perform more difficult welds than their production counterparts. It is important that maintenance workers who will be assigned welding duties know the potential risks of welding so they can protect themselves adequately and avoid dangerous situations. They should understand that fumes, eye injury, radiation, fires, and explosions are hazards the welder may encounter. They must be made aware of the equipment available to protect against these hazards and how this equipment should be used. These devices include spot fume eliminators to exhaust the fume plume before it hits the welder, welding helmets equipped with an appropriate filter plate, and fire resistant leather gloves and jacket.

The in-plant training program is made up of specialized classes that bring selected people up to a certain level of education and experience in a maintenance skill required by the company. It is important that those selected be pretested for the training to be received. Thus it is possible to tailor training to meet specific requirements rather than to provide a mass of material that will overwhelm some of the participants and bore others. The in-plant continuing education program is used often to broaden the capabilities of the personnel. By cross-training employees in multiple skills, the company will help assure quality maintenance on today's exotic facilities and the maintenance technicians will be upgraded in terms of their ability to render service.

The size of an individual group receiving this instruction may be as small as three to four or as large as 10 to 12. The class size is determined by the anticipated need based upon the maintenance technician's forecast.

The instruction is done in the plant, usually during the workday, during blocks of time of 2 to 4 hr. The remainder of the workday, for those receiving the instruction, is utilized in performing their usual assigned work. When new employees are involved, they are given work that they are able to perform with their current skill level. The length of the course usually runs from 200 to 600 hr of classroom and laboratory work.

The instruction is usually provided by some member of the plant engineering staff who is competent in the area of work and has teaching capability. At times, it may be necessary to employ an expert from the outside to handle the teaching and practical training. This is the best way to proceed when the company does not have anyone with sufficient time or skills and experience to do a quality job of training in the area. Often manufacturers of the equipment, where the majority of the maintenance effort will be utilized, are in a position to provide the necessary instruction. Sometimes it is advantageous to use a professional educator who will plan the course, handle much of the technical instruction but utilize skilled technicians to provide the laboratory instruction. These skilled technicians may be provided by the equipment manufacturer or may be company employees.

To help assure the success of the in-plant training for maintenance technicians, a well-equipped classroom should be available that accommodates up to the maximum size class being offered. There should be ample blackboard space, table-arm chairs or tables where the trainee can be seated comfortably while taking notes. The classroom should be well-lit and air conditioned. It should be kept bright and clean. At one end there should be storage cabinets where lecture handouts, appropriate books, and visual aid equipment can be stored safely. The visual aid equipment that should be available includes

1. VCR and monitor
2. 16-mm sound projector
3. 35-mm slide projector
4. Overhead projector

5. Suitable screen
6. Tape recorder

Closed-circuit television is now being effectively used to bring field installations and activities live into the classroom. Also, it is good practice to videotape training sessions of a highly technical content for subsequent viewing by students; this provides valuable instruction resources for new employees and anyone who have missed the primary training session.

It is good practice to require the trainee pass a written and practical examination as an indication of successful completion of the course. Those employees who successfully complete the course should be acknowledged with an appropriate certificate suitable for framing. A certificate should indicate the skill area that comprised the instruction, the number of lecture and practical hours involved, the company's name, the name of the trainee, the date, and the name and signature of the instructor.

PROGRAMMED INSTRUCTION

Programmed instruction is one of the best methods to provide technical information to maintenance employees in connection with training to prepare a worker for a new job opportunity. This method competes with classroom instruction from the standpoint of information retained per hour of time invested by the employee. It is also usually less costly.

Typically, in programmed instruction, pertinent subject matter is presented to the trainee in small increments. As soon as the students reads and contemplates an increment of information, they test themselves by responding to specific questions. The questions asked have been designed carefully to evaluate the understanding of the information presented. Immediately after answering a given question, the students can refer to an answer sheet to reinforce the understanding and solution of the question or problem.

The principal advantage of programmed instruction is that the maintenance workers can learn at their own pace, during their own time, and at their selected place of study. Since they are continually answering questions and reinforcing their grasp of new knowledge, the trainees' interest tends to be high. Seldom do the trainees get bored or discouraged with the presentation of each increment of information. Programmed learning does require some active participation on the part of the trainees. This may involve solving problems, answering questions, or choosing from several alternatives.

Frequently, a kit of tools and hardware will accompany a programmed instruction kit. The trainees will be given the opportunity to complete experiments and assemble an instrument on a step-by-step basis. This type of hands-on instruction will help them understand some of the theory learned in the programmed instruction. For example, in the study of elementary electronics, kits allowing experiments will reinforce the trainees' understanding of capacitance, vacuum tubes, amplifier circuits, and transistors.

APPRENTICE TRAINING

As mentioned earlier, the personnel for specialized in-plant training for maintenance technicians are people who have had apprentice training or the equivalent in a skill area fundamental to the maintenance job for which the personnel are being trained. Apprentice programs are expensive to operate and, consequently, few companies today have full-scale apprentice programs. Those craftsmen who have completed a formal apprentice training program in such areas as industrial instrumentation, machine repair, industrial hydraulics, welding, millwrighting, plumbing pipe fitting, and industrial wiring are usually excellent candidates for work in maintenance departments and often receive in-plant training in connection with specialized maintenance work.

Applicants for apprentice programs should be screened carefully. Successful candidates should pass qualifying tests and be approved by an interviewing committee comprised of representatives of both the company and the union. The program typically requires about 8000 hr in the shop working with skilled tradesmen under the general direction of an approved instructor. About 7% of the time, typically 600 hr are spent on laboratory and classroom instruction in such areas as mathematics, engineering graphics reading, electricity and electronics, pneumatics, fluidics, metallurgy, physics, and effective speech. Any company desiring assistance in establishing an apprentice program can receive assistance from the U.S. Department of Labor, Bureau of Apprenticeship and Training.

SETTING UP A MAINTENANCE APPRENTICESHIP PROGRAM

As demands for maintenance employees with specialized skills continue to grow, a company may find that it is not only desirable, but necessary, to establish an apprenticeship program to help assure an adequate supply of highly skilled service technicians. Apprenticeship is a method of training workers (1) on the job, (2) for a specific length of time, (3) at predetermined rates of training pay, (4) for a specified range of skills, (5) in one particular occupation, (6) with stated hours of classroom instruction, (7) under a written agreement between the company and the trainee.

The company must recognize that it will be necessary for them to invest considerable time and capital in the program if it is to be successful. However, this investment is usually cost-effective if there is a shortage of workers who have unique skills needed by the company.

Those engaged as trainees are employees of the company and are subject to the same rules and policies governing other employees in the firm or organization. A company may establish a program for as few as one apprentice or as many as several hundred or more, depending upon the need. The program can involve only one occupation, such as maintenance mechanic repairman, or can include several, such as carpenter, machinist, plumber/pipefitter, painter/ decorator, and so on.

The U.S. Department of Labor identifies two courses of action by the employer in starting an apprenticeship program. These depend on whether or not the employees are organized. If there is no labor agreement, the following six steps would take place (usually, the company personnel or training director would spearhead the program):

1. Organize an apprentice advisory group, made up of the production manager, at least one foreman, and at least one skilled worker.
2. Determine all the knowledge and skills needed for the occupation or occupations to be included in the program.
3. Secure the cooperation of the workers and foremen who will be expected to provide the apprentices with the direction and supervision on the job.
4. Have the advisory group visit the local vocational education director or school superintendent to arrange for necessary related classroom instruction.
5. Appoint an apprenticeship director to maintain the standards of training prescribed by the committee for the occupations involved, length of training, selection procedures, wages, tests, number to be trained, and so on.
6. Basic details of this program should be written up as a set of apprentice-ship standards.

In the event the company has a labor agreement, then the following steps would be followed:

1. Discuss the proposed program with the appropriate union official if the training involves employees who would be covered under the collective bargaining agreement.
2. Set up a joint apprenticeship committee with the union. This committee will have the responsibility of administering the program. The committee should have equal representation of labor and management, perhaps three from each.
3. The committee will arrange for necessary related classroom instruction with the local school system, usually through the vocational education school director.
4. The committee should agree on a set of standards for training, including occupations, length of training, selection procedures, wages, tests, and number of apprentices.
5. Basic details should be written up and approved as the standards of the apprentice program.
6. If the union has no interest in the specific apprenticeship plan, the company should obtain a waiver from the union so that it can adopt the alternate course of action.

There are basic standards for a good apprenticeship program that have been established by the Federal Committee on Apprenticeship. In order to conform to these minimum standards, an apprenticeship program should contain provisions for the following:

1. The starting age of an apprentice should be at least 16 years.
2. There should be a full and fair opportunity to apply for apprenticeship.
3. Selection of apprentices should be on the basis of qualifications alone.
4. There should be a schedule of work processes in which an apprentice is to receive training and experience on the job.
5. Organized instruction designed to provide the apprentice with knowledge in technical subjects related to his trade should be established. (A minimum of 144 hr per year is normally considered necessary.)
6. There should be a progressively increasing schedule of wages.
7. There should be proper supervision of on-the-job training with adequate facilities to train apprentices.
8. There should be periodic evaluations of the apprentice's progress, both in job performance and related instruction.
9. Appropriate records should be maintained.
10. There should be continuous employee–employer cooperation.

11. There should be suitable recognition for successful completion.
12. There should be nondiscriminatory practices in all phases of apprenticeship employment and training.

Supplementing the preceding 12 points, the U.S. Department of Labor has developed the following guide to help summarize what should be part of the provisions in any apprenticeship program.

1. *Occupations.* Determine what occupations or types of jobs will be covered by the program. It could be an occupation or a trade from the basic list such as maintenance mechanic repairperson or it could be a pioneering apprenticeship endeavor such as robot repairperson.
2. *Work Processes.* List the major on-the-job training processes for each occupation separately. Question whether or not these processes will develop the well-rounded skills needed. An example of a portion of a schedule of work processes for heavy-duty automotive repairperson is shown in Table 15.1.
3. *Allocation of work training time.* Determine the relative difficulty and importance of each work process and allocate the amount of training time (i.e., the time the apprentice is expected to work on the particular process or machine to become proficient).
4. *Term of apprenticeship.* In most traditionally apprenticed occupations the term of apprenticeship is well recognized. If you do not know what the term of apprenticeship should be, and you do not know of a standard practice for the occupation, list the work processes and set down opposite each process the amount of time it is agreed should be appropriate for each one. When everyone is satisfied about the time, total the hours and convert into months and years. This should give you a fairly accurate idea of the time required.
5. *Trainee qualifications.* What qualifications will applicants need to enter your program? These should be clear and objective; equal opportunity should be stressed. Is citizenship a requirement? Will they need education beyond high school? What about age limitations? Will there be a need to establish minimums and maximums? Any special physical, mental, or health requirements because of the occupation involved?
6. *Related classroom instruction.* As apprenticeship is most suited to jobs requiring broad skills and knowledge learned best on the job, apprentices will need classroom instruction related to skills. The federal committee has recommended a minimum of 144 hr a year. This is predicated on 2 hr per night twice a week during a 36 week

Table 15.1   Schedule of Work Processes for Heavy-Duty Automotive
Repairperson

Skills	Approximate hours
I. Cleaning and inspecting the parts of all types of equipment	400
II. Cylinder heads	
A. Checking and inspecting heads	20
B. Replacing valve guides	100
C. Removing and replacing valve seats	80
D. Reaming valve guides	50
E. Grinding valve seats with hard-seat grinder	100
F. Lapping valves	100
G. Checking valves with dial indicator	50
H. Installing injector tubes	50
I. Replace Welsh plugs and water test head	80
J. Rebushing rocker-arms and reaming bushings	100
K. Cleaning and replacing rocker-arm rollers	80
L. Torquing cylinder head bolts	20
M. Use of compound on head gaskets	40
N. Torquing injectors and adjustments	20
III. Cylinder blocks and liners	
A. Removing and installing cylinder sleeves	100
B. Cleaning and checking water passages	80
C. Checking counterbores for sleeves	60
D. Recutting and straightening counterbores	80
IV. Welding	
A. Acetylene-cutting, brazing, and welding	200
B. Electric-cutting and welding	300
V. Repair and maintenance of self-propelled and stationary equipment exclusive of engines	
A. Use of proper oils, greases, tools and shop equipment	100
B. Maintenance and repair of the various types of equipment used by the industry	500

Total hours: 6000

school year, usually provided by the local public vocational school
without charge. It is suggested that the local vocational school coordi-
nator be asked to assist. If there is no local vocational school in the
community, contact the school director of vocational education at the
state capitol for assistance.

7. *Number of apprentices.* The number of apprentices to be trained is determined usually by a ratio of apprentices to skilled workers (journeymen). Such a ratio is based upon the facilities available for employing and training apprentices and on future employment opportunities. Since apprentices learn from the journeymen, the quality of training largely depends on the number of journeymen available to instruct the apprentices and the ability of the journeymen as instructors. Seldom is a ratio of more than one apprentice to three journeymen feasible or effective.

8. *Apprenticeship wages.* A common method of expressing the apprenticeship wage, or at least of arriving at it, is using a percentage of the skilled worker's rate. There should be a progressively increasing schedule of wages, with increases at least every 6 months. The increases should be scheduled throughout the apprenticeship to provide both a monetary incentive and a reward for steady progress on the job. During the last period of the apprenticeship, the apprentice should receive 85 to 90% of the rate paid a skilled worker in the occupation.

9. *Supervision of apprentices.* Apprentices are customarily under the immediate instruction and supervision of the skilled worker to whom they have been assigned, and under general supervision of the appropriate foreman. In large apprenticeship programs, an apprentice supervisor is designated or employed on a part- or full-time basis and assigned the responsibility for carrying out the program. In small programs, the responsibility is basically that of the employer or a deputy.

10. *Apprenticeship agreement.* Your program should provide for the signing of an agreement of apprenticeship between each apprentice and the proper officer of the establishment for registration purposes with the appropriate state or federal apprenticeship agencies serving the area where the program is established. The agreement should contain (a) home address and date of birth of the apprentice; (b) name of the employer; (c) term of apprenticeship; (d) wage schedule; (e) length of probationary period; (f) an outline of the work process schedule; (g) number of hours per year the apprentice agrees to attend classes, subjects to be taken, and name of school; (h) any special provisions such as credit allowed for previous experience; and (i) signatures of the employer and apprentice. If a union is involved, its approval is necessary; or if a joint apprenticeship committee exists, the agreement would be approved by such a committee. Figure 15.1 illustrates an apprenticeship agreement.

The employer and apprentice whose signatures appear below agree to these terms of apprenticeship.

The employer agrees to the non-discriminatory selection and training of apprentices in accordance with the Equal Opportunity Standards stated in Section 30.3 of Title 29, Code of Federal Regulations, Page 30; and in accordance with the terms and conditions of the _____ (Name of Apprenticeship Standards) which are made a part of this agreement.

The apprentice agrees to apply himself or herself diligently and faithfully to learning the trade in accordance with this agreement.

Trade _____ Term of apprenticeship _____

Probationary period _____ Credit for previous experience _____

Term remaining _____ Date the apprenticeship begins _____

This agreement may be terminated by mutual consent of the parties, citing cause(s), with notification to the Registration Agency.

Signature of Apprentice	Name of Employer-Company
Social Security Number	Address
Address	
Apprentice's Birthday	Signature of Authorized Official
Parent or Guardian	
Approved By _____	Joint Apprenticeship Committee
Date _____	By _____
Registered By _____	Date _____

Figure 15.1 Apprenticeship agreement between apprentice and employer.

The training program should be established on the basis of what the apprentice must do and must know in order to perform the operations of the job in a safe and satisfactory manner.

Selecting the skilled workers who will conduct the training and acquainting them with the job to be done is an important first step.

They should be thoroughly skilled in those phases of the occupation they are going to teach. They should definitely be interested in the progress of apprentices on the job. In teaching skills, the job instructor will ordinarily proceed as follows:

1. Question the apprentice as to extent of knowledge about the operation or process.
2. Demonstrate each operation by slowly performing each new process or step, emphasizing key points and safety precautions.
3. Have the apprentice perform the operation, assisting if necessary.
4. Have the apprentice repeat the work several times under observation until satisfied that the apprentice can do it alone, safely, and well.
5. Continue the supervision and encourage questions to assure that the apprentice is carrying out the methods and processes that have been taught.

VOCATIONAL TECHNICAL SCHOOLS

Perhaps the greatest sources of competent employees for performing maintenance work are those high schools and post–high school programs that provide instruction and training in several areas related to the maintenance function. For example, many community colleges offer 2-year certificate programs in such areas as welding, small gasoline engine repair, electronics technology, service and operation of heavy machinery, plumbing and heating, machining, and so on. The graduates of these programs usually develop into first-class maintenance technicians. For example, one 2-year certificate program that emphasizes both theory and manipulative skills in welding stresses developments and techniques in electric, oxyacetylene, and inert gas shielded methods of welding. This background fits very well with those maintenance technicians who are responsible for diversified metal joining. This program requires the following course work:*

First semester
    Acetylene welding   13 credits
    Technical mathematics   3 credits

*Courtesy: Williamsport Area Community College, Williamsport, Pennsylvania.

Second semester
    Electric welding   13 credits
    Communications   3 credits

Third semester
    Inert gas welding   13 credits
    Engineering graphics   2 credits
    Optional elective   3 credits

Fourth semester
    Advanced welding   13 credits
    Optional elective   3 credits

CONTINUING EDUCATION FOR STAFF PERSONNEL

In this rapidly developing electronic age, it is mandatory that staff personnel participate in continuing education just as maintenance section managers and craftsmen need to be updated and broadened periodically. Those staff positions associated with the maintenance engineering functions have various titles and job descriptions. The vice president of operations, manager of plant engineering, manager of production maintenance, maintenance planner, plant engineer, director of facilities planning, and facilities maintenance manager are representatives. All of these positions have some responsibility in the development, installation, and control of the basic steps characteristic of a modern maintenance management program. These basic steps include initiation of the work order, approval of the work to be done, planning and scheduling the work, the actual performance of the work, recording meaningful data related to the work, developing the costs involved and other desirable management information, updating the history of capital equipment, and developing control reports.

A background of engineering education is desirable for these positions. Most baccalaureate programs in mechanical, industrial, and electrical engineering provide the necessary base to prepare the individual to handle all aspects of the job after receiving specialized training in connection with the range of activities of the particular work assignment.

The nature of the continuing education courses that need to be taken will, of course, vary with the particular responsibilities of each staff employee. Those staff members having responsibilities for diagnostic procedures may need to be updated on process control. They may find it desirable to take a modern course in which sensors; transmitters; final control elements; control modes; and pressure-, level-, and temperature-sensing devices are discussed. Or again this same group may find it desirable to take a course in machinery health diagnostics with subject matter including diagnosis of rolling element bearings and automatic control systems and discussions of graphic systems for fracture and failure control.

The Instrument Society of America and the IEEE Reliability Society frequently offer seminars and short courses that can be of much benefit to those staff technologists associated with the engineering maintenance management function.

SUMMARY

An ongoing training program is an important component of a successful engineering maintenance management program. This is necessary because of the rapid growth of technology and the increasing sophistication of machinery and equipment. Everyone associated with the maintenance function needs to be updated periodically in order to perform their respective work assignments effectively and in a cost-conscious manner.

The training received by maintenance workers must be broad in scope since the installation, maintenance and repair of modern process equipment requires a wide range of skills and knowledge. Many of today's technicians must have a sound background of technical knowledge that has its foundation in the physical sciences. Consequently, it is important that trainees have at least average mathematical, scientific, and mechanical aptitude backgrounds. Most of the small and medium-sized plants are neither equipped nor staffed to handle their total training needs. They usually can take care of a portion of their internal needs that tend to be common for all maintenance mechanics, such as interpretation of engineering graphics, shop mathematics, basic physics, and so on. Specialized training can be obtained in an area such as instrument maintenance from the vendors who supplied the equipment. It is highly desirable to have a well-equipped classroom on the premises of the company so that classes can be provided on company time, under ideal surroundings, and with a minimum of travel for the trainees.

For those employees who have either staff or line responsibilities in the maintenance function, there needs to be a continuing effort to stay abreast of technology. Advances in detection, diagnosis, and prognosis methods for helping to schedule preventive maintenance are continually taking place. Staff engineers need to be updated periodically in such areas as sound intensity theory, electronic switching, x-ray analysis for wearmetal particles in lubricants, diagnostics and prognostics of rotating machinery, and failure analysis based on fatigue. Also, certain staff responsibilities will need to be updated on the latest thinking in connection with inventory models, development of indirect labor standards, and important ratios in connection with maintenance performance.

As maintenance requires an ever-increasing proportion of the total company budget, it continues to be fundamental to provide training in technology—the greatest prime mover in the improvement of technology.

SELECTED BIBLIOGRAPHY

Heyel, Carl. *The Encyclopedia of Management.* New York: Van Nostrand Reinhold, 1982.

Department of Labor and Industry, Commonwealth of Pennsylvania. "Pre-employment Testing for All Occupations." ES-2272, Philadelphia: Office of Employment Security, 1988.

Ross, James R. "On-the-Job-Training." In Garriel Salvendy, Ed., *Handbook of Industrial Engineering*, 2nd ed. New York: Wiley, 1992.

U.S. Department of Labor. "Apprenticeship Past and Present." Washington, D.C.: Bureau of Apprenticeship and Training, 1977.

16
Compensation for Maintenance Work

A successful engineering maintenance management system must include a method of compensation that is fair to all the employees involved in maintenance work and fair to the company management. Thus the compensation system should provide sound base rates and a reward system to those employees who consistently perform above standard (see Chapter 4).

Sound base rates assure that rates are commensurate with the local rates for similar work. They must provide adequate differentials for jobs requiring higher skills, physical effort, poor working conditions, and increased responsibilities; and they should be based upon techniques that can be explained and justified.

JOB ANALYSIS

Sound base rates are a result of job evaluation, which can be defined as a procedure for determining the relative worth of the various work assignments in an enterprise. The basis of job evaluation is job analysis, which is the procedure for making a careful appraisal of each job and then recording the details of the work so that it can be evaluated fairly by a trained analyst. Figure 16.1 illustrates an analysis of a maintenance leader mechanic (mechanical) for

JOB ANALYSIS

Job Title __Maintenance Leader—Mechanical__ Dept. __Maintenance__

Date __11/15__ Total Points __348__ Job Class __3__ Analyst __A. B. Jones__

Job Description

Basic Functions

The Maintenance Leader—Mechanical reports to the Maintenance Supervisor. Each Maintenance Leader is assigned a functional maintenance area, and is directly responsible for the orderly and efficient execution of all maintenance work in his area. He will act to maximize the effectiveness and productivity of the maintenance employees assigned to his area. He will give prompt attention to emergency jobs, and will see that accurate daily reports of the man-hours charged against pieces of equipment are submitted to accounting.

The Maintenance Leader has no specific liaison duties with production supervision except those necessary for (1) the execution of work already planned and scheduled by the Planning Leader, (2) the continuation of work in progress from the previous day and indicated on the work schedule, and (3) the arrangement for emergency work.

The Maintenance Leader assists the Maintenance Supervisor in directing and controlling all line maintenance personnel. The Maintenance Leader refers all disciplinary problems to the Maintenance Supervisor.

Responsibility and Authority

The Maintenance Leader is responsible for and has the authority to discharge the following duties:

With Respect to Work Execution

1. Carefully reviews the daily work schedule for the following day, together with the corresponding work orders, and informs maintenance employees of their work assignments for the following day.

2. Before starting a job, verifies that all predeterminable tools, equipment and materials are available, and the facility to be worked on is available.   (continued)

Figure 16.1 Sample analysis of a maintenance leader mechanic.

3. Releases jobs to work with minimum delay, and wherever possible gives the maintenance employees their next assignment before they complete the assignment that they currently are working on. These assignments are made while in the field.

4. Assists in assignments of employees to each job.

5. Sees that the maintenance personnel are fully aware of and use the best work methods.

6. Coordinates and executes all real emergency jobs.

7. Reassigns or relocates workers and/or equipment when necessary because of changes in job priorities. Refers questionable work orders or job plans to the Planning Leader for clarification.

8. Instructs maintenance personnel as to the safety procedure to be observed in performing their assigned jobs.

9. Assists in the performance of related work assignments as directed.

With Respect to Follow-Up

1. Follows up assigned scheduled jobs to assure that the work is completed within the time frames established on the work schedule. If more time is utilized, identify the reason or reasons.

2. Assures that the maintenance personnel know and understand the "what, how, where, and when" of the jobs to which they are assigned.

3. Inspects work in progress and completed jobs to assure that the work meets established quality standards.

4. Identifies and, if possible, rectifies job delays that are affecting maintenance employees productivity.

With Respect to Advisory Activities

1. As necessary, assists the Planning Leader with the planning and scheduling of maintenance work.

2. Reviews work schedules and advises the Planning Leader of any changes or revisions necessary because of unforeseen job requirements or emergencies.   (continued)

Figure 16.1 *Continued*

3. Furnishes the Planning Leader with the daily status of jobs indicating those jobs which will be completed by the end of the shift and those expected to carry over into the following day.

4. Assists in assigning manpower to schedule jobs.

5. Suggests improved methods to the Industrial Engineering Department through the Maintenance Supervisor and Planning Leader. This effort is especially important in connection with repetitive jobs.

6. Directly assists in the planning for equipment shutdowns that occur in the maintenance area.

7. Reviews drawings, specifications, plans, etc. as requested by Plant Engineering and recommends changes when necessary.

8. Reviews and examines the causes of major repairs with the Maintenance Supervisor and Planning Leader.

9. Assists in the storeroom in specifying and maintaining spare parts and other maintenance materials.

10. Reports any damage to plant and/or facilities that occurs during work execution to the affected Production Supervisor.

11. Reviews work methods, procedures, tools, equipment and manpower distribution continuously and recommends changes, where necessary, to maintain efficient operations.

12. Reviews conditions of facilities and equipment in his or her area and submits periodic reports to the Maintenance Supervisor.

Job Evaluation	Degree	Points
Education	3	42
Experience and training	5	110
Initiative and ingenuity	4	56
Physical demand	1	10
Mental and/or visual demand	3	15
Responsibility for equipment or process	3	15
Responsibility for material or product	3	15
Responsibility for safety of others	4	20
Responsibility for work of others	4	20
Working conditions	3	30
Unavoidable hazards	3	15
	Total	348

use in a point job evaluation plan. Typically, the skills, job responsibilities, experience required, and working conditions are identified. Also the consequences resulting from poor decisions are included and information as to the tools and machines utilized should be cited. The physical and social conditions related to the job should be outlined.

JOB EVALUATION

The principal purpose of job evaluation is to determine the proper compensation for the work performed on each job. A well-conceived job evaluation plan will

1. Provide a basis for explaining to employees why one job is worth more or less than another.
2. Provide a reason to employees for adjustments of rate of pay as the content of the work assignment changes.
3. Provide a basis for assigning personnel with specific abilities to certain jobs and the selection of personnel to receive specific on-the-job training.
4. Help determine the criteria for a job when employing new personnel or making promotions.
5. Provide assistance in the training of supervisory personnel.
6. Provide a basis for determining where opportunities for method improvement exists.

Although there are four principal methods of job evaluation being practiced (classification method, factor comparison method, ranking method, and point system), only the point system will be discussed here. A well-designed point system is both objective and thorough in its evaluations of the various jobs involved. It compares all the different attributes of a job with those same attributes in other jobs.

The following procedure should be followed when installing a point system job evaluation plan:

1. Establish and define the basic factors which are common to most jobs and which indicate the elements of value in all jobs.
2. Specifically define the degrees of each factor.
3. Establish the points to be accredited to each degree of each factor.
4. Prepare a job description of each job.
5. Evaluate each job by determining the degree of each factor contained in it.

6. Sum the points for each factor to get the total points for the job.
7. Convert the job points into a wage rate.

The number of factors selected usually ranges between four and 15. When as few as four are used, they usually are (1) skill, (2) effort, (3) responsibility, and (4) job conditions. Plans that use more factors usually refine these four. For example, responsibility may be broken down into four factors: responsibility for equipment or process, responsibility for materials or product, responsibility for safety of others, responsibility for work of others. It is wise to use only as many factors as are necessary to provide clear cut differences among the jobs of the particular company. The elements of any job may be classified as to

1. What the job demands that the employee bring in the form of physical and mental factors
2. What the job takes from the employee in the form of physical and mental fatigue
3. The responsibilities that the job demands
4. The conditions under which the job is done

The selection of factors is generally the first task that is undertaken when introducing job evaluation. Those factors identified by the National Electrical Manufacturers Association (NEMA) represent a good starting point in the development of the factors to be used in a specific plant. These factors are education, experience, initiative and ingenuity, physical demand, mental or visual demand, responsibility for equipment or process, responsibility for materials or product, responsibility for safety of others, responsibility for work of others, working conditions, and hazards.

These factors usually represent all that are needed in the typical metal trade plant. The vast majority of maintenance work is evaluated satisfactorily using these factors.

All these factors are present in varying degrees in the various jobs, and any specific job under study will fall under some one of the several degrees of each factor. Usually the various factors are indicated as having different importance. In order to provide recognition to these differences in importance, points are assigned to each degree of each factor as shown in Table 16.1. Figure 16.2 illustrates the job rating and substantiating data sheet for a typical maintenance machinist based on the NEMA system.

In order to arrive at the points characterized by the job under study, each degree of each factor is defined carefully in sufficient detail so that it is evident to the trained analyst as to what degree characterizes the work situation. For

Table 16.1 Points Assigned to Factors and Key to Grades

Factors	1st degree	2nd degree	3rd degree	4th degree	5th degree
Skill					
1. Education	14	28	42	56	70
2. Experience	22	44	66	88	110
3. Initiative and ingenuity	14	28	42	56	70
Effort					
4. Physical demand	10	20	30	40	50
5. Mental and/or visual demand	5	10	15	20	25
Responsibility					
6. Equipment or process	5	10	15	20	25
7. Material or product	5	10	15	20	25
8. Safety of others	5	10	15	20	25
9. Work of others	5	10	15	20	25
Job conditions					
10. Working conditions	10	20	30	40	50
11. Unavoidable hazards	5	10	15	20	25

example, education may be defined as appraising the requirements for the use of mathematics, engineering drawings, measuring instruments, and trade knowledge. Then, perhaps, first-degree education might require only the ability to read and write and to add and subtract whole numbers. Second-degree could be defined as requiring the use of simple arithmetic, such as division and the addition and subtraction of decimals and fractions, together with simple drawings and some measuring instruments, such as calipers and a scale, characteristic of two years of high school. Third-degree education may require the use of fairly complicated drawings, advanced shop mathematics, handbook formulas, and a variety of precision measuring instruments plus some trade knowledge in a specialized field or process. It could be thought of as being equivalent to four years of high school plus short-term trade training. Fourth-degree education could require the interpretation of complicated engineering drawings and specifications, use of trigonometry and algebra, and use of a wide variety of precision instruments plus broad shop knowledge. It may be equivalent to four years of high school plus four years of formal trade training. Fifth-degree education may require a basic technical knowledge sufficient to deal with complicated electronic, electric, pneumatic, hydraulic, mechanical, or other engineering problems. Fifth-degree education

Job Rating—Substantiating Data
Dorben Mfg. Co.
State College, Pennsylvania

Job Title Maintenance machinist Code 211 Date January 15

Factors	Deg	Points	Basis of rating
Education	3	42	Requires the occasional use of fairly complicated equipment drawings, shop mathematics, variety of precision instruments, shop trade knowledge. Equivalent to four years of high school or two years of high school plus two to three years of machinist or similar trades training.
Experience	4	88	Three to five years maintaining, repairing, and installing diversified machine tools and other production equipment.
Initiative and ingenuity	3	42	Rebuild, repair, and maintain a wide variety of medium size standard, automatic, and tape controlled machine tools. Diagnose trouble, disassemble machine and fit new parts, such as bearings, spindles, gears, cams, etc. Produce from stock replacement parts if necessary. Involves skilled and precise machining using a variety of machine tools. Judgment required to diagnose and remedy trouble quickly in order to maintain production.
Physical demand	2	20	Intermittent moderate physical effort required in the tearing down, assembling, installing and maintaining machines.

(continued)

Figure 16.2 Typical job rating and substantiating data sheet.

Factors	Deg	Points	Basis of rating
Mental or visual demand	4	20	Periodic concentrated mental and visual attention required. Laying out, setting up, machining, checking, fitting parts on machines.
Responsibility for equipment or process	3	15	Damage seldom over $1000. Broken parts of machines. Carelessness in handling gears and intricate parts may cause damage.
Responsibility for material or product	2	10	Probable loss due to scrapping of materials seldom over $500.
Responsibility for safety of others	3	15	Normal safety precautions are required to prevent injury to others: fastening work properly to face plates, handling fixtures, etc.
Responsibility for work of others	2	10	Responsible for directing one or more helpers a great part of time. Depends on type of work.
Working conditions	3	30	Typical machine shop conditions. Some exposure to oil, grease, machine chips, etc.
Unavoidable hazards	3	15	Some exposure to flying particles, electric shock, and such accidents as cuts, abrasions, crushed hand or foot, etc.

Remarks: Total 307 points. Assigned to Job Class 3

Figure 16.2 *Continued*

could be thought of as being equivalent to four years of technical college or university training.

Experience appraises the length of time that an individual with the specified education usually requires to perform the work satisfactorily from the standpoint of both quality and quantity. Here, first-degree could involve up to three months; second degree, three months to one year; third degree, one to three years; fourth degree, three to five years; and fifth degree, over five years.

In a similar manner, each degree of each factor is identified with a clear definition and with specific examples when applicable.

Performing the Evaluation

Considerable judgment is exercised in the evaluation of each job with respect to the degree required of each factor utilized in the job evaluation plan. Consequently, it is usually desirable to have a committee perform the evaluation. The committee customarily is chaired by the company personnel director or someone from that office. Other members of the committee could be a union representative, the department foreman, the department steward, and someone from the industrial engineering department. A separate committee should be appointed for the evaluation of the personnel in each department of the company or business.

When meeting, the committee should evaluate all jobs for the same factor before proceeding to the next factor. For example, all jobs in the department under study (carpenter, plumber, millwright, electrician, machinist, etc.) should be evaluated for the degree of skill before proceeding to other factors, such as effort, responsibility, and job conditions. By using this pattern of procedure, the committee will measure the job rather than the particular individual filling the job.

It is important that committee members assign their degree evaluations independent of the other members. After all committee members have made their separate evaluations, all members should discuss any differences that may exist until there is agreement on the level of the factor. A meeting should not be adjourned until the factor under study has been evaluated for all jobs that are being evaluated in the department.

Classifying the Jobs

After all jobs have been evaluated, the points assigned to each job should be tabulated. The number of labor grades within the plant should now be decided. This number is a function of the range of points characteristic of the jobs within the plant. Typically, the number of grades runs from eight (typical of smaller plants and lower technology industries) to 15 (typical of larger plants and higher technology industries) (see Fig. 16.3). The number of grades that embrace maintenance hourly workers is typically four or five. For example, if the point range of all jobs within a plant ranged from 110 to 365, the grades shown in Table 16.2 could be established.

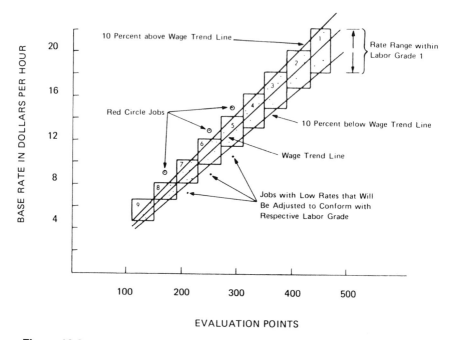

Figure 16.3 Evaluation points and base rate range for nine labor grades. (Courtesy of B. W. Niebel, *Motion and Time Study*, Richard D. Irwin, Inc., Homewood, IL.)

Table 16.2 Representative Labor Grade–Job Point Relationship

Labor grade	Range of job points
1	360 and above
2	338–359
3	316–337
4	294–315
5	272–293
6	250–271
7	228–249
8	206–227
9	184–205
10	162–183
11	140–161
12	100–139

It is not necessary to have like ranges for the various labor grades. It may be desirable to have increasing point ranges for more highly compensated jobs (see Fig. 16.3).

All the jobs falling within the various labor grades should now be reviewed in relation to one another to assure fairness and consistency. For example, it would not be appropriate for a maintenance machinist, class A to be in the same grade level as a maintenance machinist, class B.

The next step is to assign hourly rates to each of the labor grades. These rates are based upon area rates for similar work and company policy. Often a rate range is established for each labor grade. The total performance of each employee, measured regularly, will determine the pay rate within the established range. Total performance refers to quality of work, ability to exceed normal performance, suggestions, good attendance, good safety record, and so on.

Installation of the Job Evaluation Program

After area rates have been plotted against the point values of the various jobs, a rate-versus-point value trend line is plotted. This trend line may or may not take the form of a straight line. Regression techniques are recommended in establishing the trend line. After the trend line has been developed, it will be noted that on occasion several points may be above or below the trend line. Points significantly above the trend line represent employees whose present rate is higher than that established by the job evaluation plan, and points significantly below the trend line represent employees whose present rate is less than that prescribed by the plan.

Employees whose rates are less than that called for by the plan should receive an immediate increase to a new base rate established by the job evaluation plan. Employees whose rates are higher than that called for by the plan (such rates are referred to as *red circle rates*) are not given a rate decrease. They are, however, not given an increase at the next contract review unless the cost of living adjustment results in a rate higher than their current pay. Any new employee would be paid the new base rate established by the job evaluation plan.

Salaried Point Job Evaluation Plans

Just as it is important to determine the proper compensation for the work performed by hourly workers, so it is desirable to reward equitably salaried workers for their various work assignments. Salaried personnel associated with maintenance include office, clerical, technical, supervisory, and managerial personnel. Specific titles include plant engineer, maintenance supervisor, planning leader, maintenance leader—mechanical, maintenance leader—electrical, and so on.

When introducing a salaried job point evaluation plan, the first step is to assure that well-prepared individual job descriptions exist. If they do not exist, they should be prepared by a representative from the industrial relations department as follows:

1. Schedule a preliminary discussion with the department head and the incumbent's immediate supervisor; obtaining approval for the development of an evaluation for the job in question.

2. Schedule an interview with the incumbent wherein a record is made of the incumbent's description of the duties and responsibilities. During this interview, the incumbent should be queried as to specific aspects of the job and how the incumbent integrates with other people in the organization both vertically and horizontally.

3. Schedule an interview with the incumbent's immediate supervisor to determine the limits of authority and responsibility relative to the position. Also determine the comparative position of the job in question with other jobs in the department.

4. Make an analysis of the data obtained. This analysis will utilize notes from the interview with the incumbent, notes from the interview with the supervisor, and related job descriptions.

5. Prepare a description draft with a view to:
   a. Obtaining accuracy of descriptions on the basis of responsibilities, not personalities
   b. Establishing the scope and limits of authority and responsibilities
   c. Analyzing and recommending revisions to position contents where unclear or overlapping areas exist
   d. Including of position guides and goals
   e. Providing functional descriptions which can be used by the supervisor and incumbents in establishing operational standards, development and training programs, and reviewing the organization structure
   f. Establishing a basis for detailed methods and procedures

6. Review the draft with the incumbent. At this time, the analyst will discuss interpretations of the original interview. Necessary revisions will be introduced.

7. The draft will now be reviewed with the incumbent's supervisor and necessary revisions will be incorporated.

8. The education, experience, and other requirements characterized by the point plan being used will now be applied.
9. The job description will now be evaluated by the analyst based upon those factors being used in the point plan.
10. The job description and evaluation will be reviewed by the industrial relations manager or a representative. This review may result in some necessary revisions.
11. Finally, the revised draft will be reviewed with the incumbent, the supervisor, and the department head. Approval will be obtained by signature from these three people.

After all salaried job descriptions within a department (maintenance or other) are completed, the analyst should note crossover references to other departments (for example, the relation of maintenance staff to production staff). The analyst should also make a vertical analysis noting the existing and proposed grades of all salaried employees in the department.

When all descriptions companywide have been completed, following the procedure described above, a horizontal analysis should be made. This analysis may result in some revisions. The job description of a maintenance supervisor in Fig. 16.4 is indicative of the amount of detail that should be included.

With the completion of all salaried job descriptions in the plant, evaluations can be made. As in hourly evaluation plans, salary evaluation is done, usually by a salary administration committee chaired by the manager of industrial relations. The committee should also include the chief industrial engineer and a representative of top management. The factors to be evaluated generally will be somewhat different from those used in hourly plans. Typically a salary plan will identify various levels of the following factors for each salaried position: education, experience, initiative and judgment, analytical ability, personal influence, supervisory responsibility, monetary responsibility. It may be desirable to consider physical application, metal or visual application, and working conditions.

The number of job classes in a salaried plan will vary depending upon the size of the company or business and the total coverage of the plan. For example, top management positions may not fall under the plan. Here, a profile ranking procedure based upon the job descriptions may be used to determine the salaries of these administrators.

If the plan extends from clerical to top management there would be approximately twelve job classes with perhaps as many as ten grades within each class. It should be understood that the plotting of salary-versus-grades is not necessarily linear. In fact it is typical for them to be nonlinear (see Fig. 16.5).

Issued 10/1	Revised	**Armstrong**	Section III	Page 5
Approved		Armstrong World Industries, Inc. SOUTH GATE PLANT MAINTENANCE GUIDE		

JOB DESCRIPTION

MAINTENANCE SUPERVISOR

Basic Functions

The Maintenance Supervisor reports directly to the Plant Engineer in all phases of
his work. He is directly in charge of all maintenance work in the plant.

His prime function is to provide effective supervision and coordination of all
maintenance work. He is responsible for developing a smooth, efficient team
so that maintenance work is performed efficiently, safely, and on time with
minimum expenditures of manpower, materials, tools, supplies and utilities.

The Maintenance Supervisor is responsible for the personnel administration of all
maintenance employees. That is, he is responsible for certain personnel records,
training, transfers, reclassifications, and the administration of Company
policies pertaining to vacations, sickness, absences, performance, etc. He is
responsible for all disciplinary measures.

The Maintenance Supervisor must be in constant communication with his Maintenance
Leaders. This communication is direct and informal in every respect with information flowing in both directions so that decisions can be made without loss
of productivity of people and/or equipment.

The entire maintenance work force is responsible to the Maintenance Supervisor.
The maintenance employees are subject to direct orders from the Maintenance
Supervisor in matters concerning the following:

            1. Execution of Work
            2. Deployment of Personnel
            3. Quality of Work
            4. Productivity
            5. Housekeeping
            6. Methods of Doing Work
            7. Tools and Their Proper Use
            8. Safe Practices
            9. Coordination of Work
           10. Discipline
           11. Utilization of Personnel and Materials

Responsibility and Authority

The Maintenance Supervisor is responsible, and has the authority to discharge, the
following duties:

With Respect to Work Execution

1. Carefully reviews the daily work schedule for the following day, together
    with the corresponding work order, and working with the Maintenance Leaders
    assigns maintenance employees their work scheduled for the following day.

Form 43633 6/80J

(continued)

Figure 16.4   Sample job description for maintenance supervisor.

Maintenance Work Compensation

Issued 10/1	Revised	**Armstrong**	Section III	Page 6
Approved		Armstrong World Industries, Inc. SOUTH GATE PLANT MAINTENANCE GUIDE		

2. Coordinates and executes, in an efficient manner, all real emergency jobs.

3. Makes frequent daily personal surveys of the maintenance work areas to:

   a. Follow up the scheduled jobs with each Maintenance Leader to assure that the work is completed within the time frames established on the work schedule...and if not, why not.

   b. Detect deviations from work schedules (nonscheduled work) and examine the causes. Report trouble spots to the Planning Leader, and/or the Plant Engineer for remedial action.

   c. Assure that the Maintenance Leaders are using maintenance employees in the most efficient manner...that the work pace is steady and time losses minimized, and that the maintenance employees stay on the job.

   d. Assure that good housekeeping practices are followed and that the safety rules are enforced.

   e. Observe the progress of the work to make sure that the important jobs are done first (as conditions may change) and that all work is completed on time.

   f. Satisfy himself that his Leaders are exercising control over jobs... and if not, why not.

   g. See that preventive maintenance work is not slighted.

   h. See that completed work is checked by his Leaders to assure high quality work.

   i. Make sure that preparations have been made for the next day's work.

4. Makes certain that policies are understood and procedures regarding discipline and work rules are carried out.

5. Provides a source of technical advice for his Leaders.

6. Makes certain that work orders, daily work schedules, and time distribution are factually correct and on time.

7. Inspects maintenance facilities and equipment for continued serviceability, obsolescense, and/or the need for work orders.

8. Makes certain that the shops are neat and in an orderly condition.

9. Coordinates the activities of independent contractors working conjunctively with the maintenance force on projects.

10. Inspects the work performed by outside contractors to assure compliance with contract provisions and/or specifications, and to assure that their work meets proper quality standards.

Form 43633 6/80J

(continued)

Issued 10/1	Revised	**(A)rmstrong**	Section III	Page 7
Approved		Armstrong World Industries, Inc. SOUTH GATE PLANT MAINTENANCE GUIDE		

With Respect to Advisory Activities

1. Assists the Planning Leader to update the preventive and corrective maintenance program.

2. Assists the Planning and Maintenance Leaders to analyze equipment failures.

3. Assists the Planning Leader to develop better plans, estimates and schedules.

4. Makes appraisals on whether the work involved on work orders should be executed by plant personnel or by an outside contractor. Obtains the Plant Engineer's approval for the work to be performed by outside contractors. The Plant Project Engineer will handle the larger jobs to be contracted, and obtain any additional approval required. The smaller jobs that require no preapproval will be delegated to the Planning Leader and he will make the necessary arrangements with the contractor.

5. Schedules the Shift Mechanics.

6. Reviews shutdown schedules affecting production and/or maintenance conditions with the Planning Leader.

7. Reviews emergency jobs and downtime statistics, and discusses causes and possible improvements with the Planning Leader and Maintenance Leaders and the affected production supervision.

8. Provides the necessary follow-through on work schedules, manpower availability and other communications for night, holiday and weekend shifts.

9. Reviews the inventory of spare parts and materials, periodically, to insure an adequate and proper inventory is maintained.

10. Assists Plant Engineering by:

    a. Recommending changes and renovations to existing facilities and equipment.

    b. Aiding in developing the designs for new equipment and facilities.

With Respect to Departmental Development

1. Periodically reviews the performance of the maintenance employees and counsels them about improvement.

2. Institutes training programs (in consultation with Personnel and others) for craft training, work methods changes, cost improvement, etc.

Form 43633 6/80J

(continued)

Figure 16.4 *Continued*

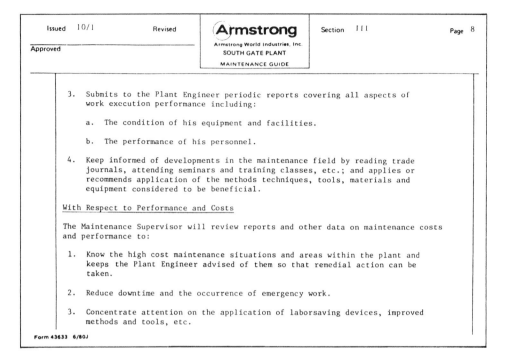

Armstrong World Industries, Inc.
SOUTH GATE PLANT
MAINTENANCE GUIDE

3. Submits to the Plant Engineer periodic reports covering all aspects of
work execution performance including:

   a. The condition of his equipment and facilities.

   b. The performance of his personnel.

4. Keep informed of developments in the maintenance field by reading trade
journals, attending seminars and training classes, etc.; and applies or
recommends application of the methods techniques, tools, materials and
equipment considered to be beneficial.

With Respect to Performance and Costs

The Maintenance Supervisor will review reports and other data on maintenance costs
and performance to:

1. Know the high cost maintenance situations and areas within the plant and
keeps the Plant Engineer advised of them so that remedial action can be
taken.

2. Reduce downtime and the occurrence of emergency work.

3. Concentrate attention on the application of laborsaving devices, improved
methods and tools, etc.

Form 43633 6/80J

Promotion within job classes is made progressively up the scale of the labor
grades within each class, which are usually arranged in arithmetic progression.

It is a good idea to have both the incumbent's immediate supervisor and
the division head review the evaluation made by the salary administration
committee.

Like any job evaluation plan, in order to assure its continued success, it is
important that it receives the active and complete support of top management
and that it be audited periodically.

THE COMPENSATION PLAN

Certainly the vast majority of maintenance and other indirect workers in
American industry and business today are compensated on an hourly basis.
The hourly base rate, as we have learned, should be established by job
evaluation. Only about 25% of manufacturing employees in total are compen-

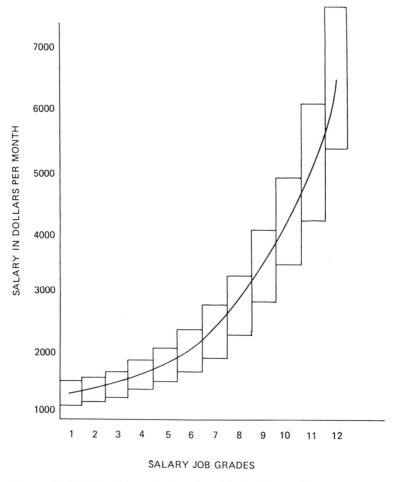

Figure 16.5   Salary job evaluation plan with 12 salary grades.

sated on the basis of some incentive plan. The principal industries that utilize incentives heavily throughout the plant are the textile and basic steel industries.

In this text, incentive wage payment is being advocated for hourly workers (both direct and indirect) for two reasons:

1. The typical employee will not do an extraordinary day's work for an ordinary day's pay. In today's competitive economy, it is important that all employees continually produce at more than normal performance.

2. With fringe benefits becoming increasingly significant (they average about 30% of the basic wage), it is important that these costs be spread over more production and maintenance. Fringe benefits average approximately 19 vacation days a year, 9 paid holidays a year, $13,000 in life insurance, disability insurance at 72% of salary, and sick pay up to 85 days.

All incentive plans that result in an increase in the employees production will fall under one of the following three classes: direct financial plans, indirect financial plans, and plans other than financial.

Direct financial plans include all plans in which the maintenance worker's compensation is commensurate with output. In this category are included both individual incentive plans and group plans. In the individual plan, each maintenance employee's compensation is governed by performance for the period in question. Group plans are applicable to two or more persons who are working as a team on operations that tend to be dependent on one another. In these plans, each employee's compensation within the group is based upon a base rate and on the performance of the group for the period in question.

Of course, both individual and group plans may be combined in the typical maintenance operation where crew sizes can range from one to several workers. A given worker assigned a work order to be completed by one's self can be compensated on an individual basis; and then, later in the week, that worker may be part of a crew which will be compensated on the basis of the group's performance.

The incentive for increased and prolonged individual effort is greater in individual plans than in group plans. Therefore, the individual incentive plan should be given perference over group systems.

Indirect financial plans refer to those company policies that stimulate employee morale resulting in good employee performance, yet have not been designed to bring about a direct relation between the amount of compensation and the amount of productivity. Such overall company policies as fair and relatively high base rates, equitable promotion practices, sound suggestion systems, a guaranteed annual wage, and relatively high fringe benefits lead to building healthy employee attitudes which stimulate and increase productivity. Thus these policies may be thought of as indirect financial plans.

The reader should be cautioned that all indirect incentive methods have the weakness of allowing too broad a gap between employee benefits and productivity. After a period of time, employees take for granted the benefits bestowed upon them and fail to realize that the means for their continuance must result from the employees' productivity.

Plans other than financial incentives include any rewards that have no relation to pay, and yet improve the spirit of the employee to such an extent that added effort and performance result. Under this category come such company policies as periodic shop conferences, quality control circles, frequent talks between the supervisor and the employee, job enlargement, job enrichment, and several other techniques utilized by progressive managements.

The compensation plan recommended here embraces all three classes: direct, indirect, and plans other than financial. It incorporates the utilization of quality control circles, continual on-the-job training to promote job enlargement and job enrichment, and frequent discussion between line supervision and employees. These company policies provide the basis of good employee morale and high performance.

Base rates established by a point job evaluation plan, where the trend line is plotted at a level slightly higher than the mean rate paid for similar work in the area, are recommended. This policy will soon give the company or business an image of being a good organization with which to be associated. In time, the overall quality of employees will be better than the average employee in the area and will result in higher productivity.

Finally, a standard hour plan of compensation is recommended. Here the maintenance workers are rewarded throughout in direct proportion to their output. Standards are established for all work orders and the employees are compensated for this amount of time regardless of how long it actually took to perform the work. Even if it took longer to perform the work than allowed by the standard, the employee will be compensated on the basis of the time required to complete the job (see Figure 16.6). Thus the base rate (set by a point job evaluation plan) is always guaranteed. An example for the compensation of a maintenance electrician over a period of one week follows. We will assume the base rate to be $16.25 an hour. Work completed over the week is as follows:

Job	Standard	Time taken
a	14.5 hr	10.8 hr
b	4.0 hr	4.2 hr
c	1 hr	0.8 hr
d	20.3 hr	18.9 hr
e	6 hr	5.3 hr
Total	45.8 hr	40.0 hr

Compensation for work $= (\$16.25)(45.8) = \$744.25$
Performance for week $= 45.8/40 = 114.5\%$
Earned hourly rate for week $= \$744.25/40 = \$18.60$

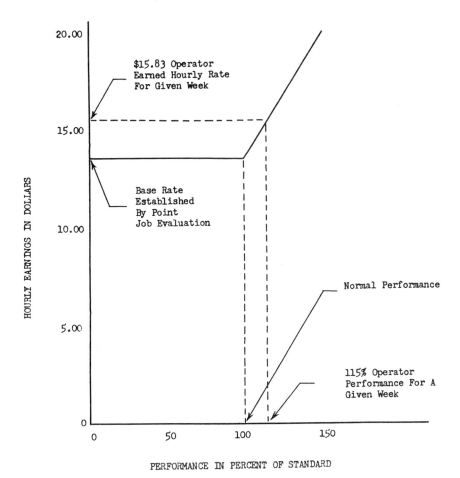

Figure 16.6 One-for-one incentive plan where worker is rewarded in direct proportion to output above standard.

SUMMARY

In order to help assure the success of the engineered maintenance management system, there should be a reward system based upon what the employee brings to the company in the form of specific skills and experience and what the employee's total productivity is during a given period of time. For hourly employees, a point job evaluation plan is recommended for establishing an equitable base rate. The talented and highly productive hourly employee is rewarded in two ways. The base rate can increase within the job class, and the compensation can be in proportion to output with a carefully designed incentive plan.

Of course, as an employees' skills and experience increase, they can be promoted to a higher class and then improve their position still further by being totally productive in this new job.

Job evaluation is also strongly recommended for establishing salary schedules for clerical and managerial positions. Salaried employees should have their performance evaluated at least once every year. Merit rating should be done by each incumbent's supervisor with the assistance of a representative from the personnel department. All merit rating should be done in writing and must be substantiated by evidence of employee growth and increased productivity to the company or business.

The successful compensation plans in today's economy are more concerned with the employee's total contribution and less concerned by regular annual adjustments to reflect the change in the cost of living index and the compensation being paid by other industries and businesses.

SELECTED BIBLIOGRAPHY

Niebel, Benjamin W. *Motion and Time Study*. Homewood, IL: Richard D. Irwin, 1982.

Salvendy, Gabriel, and Seymour, Douglas W. *Prediction and Development of Industrial Work Performance*. New York: Wiley, 1973.

Zollitsch, Herbert G., and Langsner, Adolph. *Wage and Salary Administration*, 2nd ed. Cincinnati: South-Western Publishing Co., 1970.

17

Measuring and Improving Maintenance Performance

It has been stated that the overall objective of every maintenance program should be to make the greatest possible contribution to the long-term profitability of the company. To maximize this contribution, it is necessary to know where the organization currently stands and what improvement possibilities exist. Measures of performance need to be made and evaluated. Improvement should be taking place. For the overall engineered maintenance management program to have high performance, there must be high performance in the 11 ingredients that comprise the system. These ingredients are organization, control systems, estimating and measurement of maintenance work, inventory control and use of maintenance materials, planning and scheduling, predictive and preventive maintenance, diagnostic techniques, application of the digital computer, training of maintenance employees, compensation of maintenance employees, and reports to management.

ORGANIZATION

In order to evaluate a maintenance organization, the following points should be considered:

1. The organization needs to be staffed with competent line supervision (foremen) who have adequate time to supervise effectively the skilled

tradesmen reporting to them. This usually is between 12 and 20 trades-men per supervisor.

2. The functions and responsibilities of plant engineering and mainte-nance should be clearly and thoroughly written and be endorsed by top management. The plant engineer, who should have the responsibil-ity of maintenance, should report to the top officer of the plant—usually, the plant manager. The plant engineer should not report to the production manager.

3. The foremen who supervise the skilled tradesmen should supervise a variety of trades with responsibility for a specific plant area. This procedure has proven to be more cost-effective than supervising one trade only, with plant-wide responsibility.

4. To assist in an ongoing continuous quality improvement effort, there should be an in-house benchmarking service that identifies the best maintenance practices in at least 15 different manufacturing firms. Some of the programs' activities include visiting plant sites to share best maintenance practices from other locations and learning more about a particular plant's procedures and philosophies regarding pre-dictive maintenance, preventive maintenance, routine maintenance, and subcontracting maintenance. Also, the application of job enlarge-ment on the part of production workers to incorporate preventive maintenance in their job descriptions should be surveyed.

CONTROL SYSTEMS

In connection with control systems, the following points should be considered with reference to cost control:

1. Written requests or orders must be completed for all maintenance work done.

2. Material requisitions must be used to both control and compile material costs.

3. Items and services should be purchased when they cannot be produced economically or performed by the plant facilities and personnel.

4. There should be a record of improvement in total maintenance costs over the past year.

5. The maintenance expenditures in the production cost centers should have diminished in the past year.

6. Maintenance expenditures as a percentage of sales should be improving constantly.

MEASUREMENT OF MAINTENANCE WORK

Work measurement permits the establishment of time standards for the performance of maintenance tasks. Standard data, taken from proven time studies, allows the accurate estimating of the time required to perform a task in advance of the work being done. Maintenance performance is invariably improved when it is known how long it should take to complete a given maintenance assignment instead of how long it has taken to get the job done. Another tool of the work measurement analyst is work sampling, which is perhaps the most expedient method of identifying problem areas so that maintenance work performance can be improved. Evaluation of the estimating and measurement phase of the integrated program of engineered maintenance management should recognize that

1. The approximate time required for performing each maintenance task should be known in advance of the work being begun so that effective scheduling and cost control can be accomplished.
2. Crew sizes can best be determined if reliable work standards are available.
3. A modern work standards effort will utilize several alternative techniques in the overall estimating and standards procedure. These techniques should include stopwatch time study, standard data, fundamental motion data, and universal maintenance standards. The microprocessor should be utilized in the development of standards in conjunction with all the aforementioned techniques.

INVENTORY CONTROL AND USE OF MAINTENANCE MATERIALS

Just as it is important to have the employee with the right skills available to do the work, it is important to have the right materials for a job, in the correct condition and amount. The successful maintenance materials management effort should provide for

1. Constant investigation of new materials that will provide superior performance at less cost.
2. Accepted materials that are standardized to control the quality and assure competitive prices.
3. Optimum inventories. This effort should be based upon valid analytical models that allow regular reviewing, alteration of quantities, and obsoleting inventories.

4. An on-going analysis of materials requisitions and purchase order requests in order to determine conformance to specifications, quantity in relation to optimum inventory, and availability.

PLANNING AND SCHEDULING

The cost-effectiveness of a maintenance effort is dependent to a large extent on the quality of the planning and scheduling program. Planning and scheduling should be initiated and developed by maintenance supervision with the assistance of a planning activity to provide the most effective time utilization. The following criteria related to planning and scheduling will provide measures of the quality of this effort:

1. Planning and scheduling should be based on the maintenance backlog and work priority with allowance for emergency work.
2. A full day's workload based upon standards established by measurement, developed at least one day in advance for each tradesperson, results in savings in maintenance labor through better utilization of maintenance personnel.
3. Planning and scheduling will assist in assuring standard or higher operator performance.
4. Planning and scheduling provides a means for continual analysis of the work backlog, thus triggering action when maintenance forces should be increased or decreased.
5. The use of hard output (from the microprocessor or the main frame computer or hand-posted planning boards) is desirable to maintain a perpetual record of the current status of all work assignments.
6. An assigned planning group along with maintenance supervision must initiate and develop the planning and scheduling effort.
7. The record of gross estimating errors for each craft on each job should be maintained. This record should indicate constant improvement.
8. A record of work effectiveness should be maintained and this record should indicate regular improvement.
9. A compliance record based upon the daily work schedule on both man-hours and jobs should be maintained. This performance record should indicate at least 80%.
10. The number of past-due jobs should be decreasing.

PREDICTIVE AND PREVENTIVE MAINTENANCE

Predictive and preventive maintenance is a way of life in every modern maintenance program. This effort is essential in order to minimize production downtime and maximize the effectiveness of the maintenance effort. To attain a good predictive and preventive maintenance program, the following factors should be evaluated:

1. Historical records of machinery and equipment repair are essential for the establishment of the program.
2. The predictive and preventive maintenance program should include a facility correction effort. This effort involves a modification or redesign of facilities in order to remedy deficient design from a maintenance standpoint.
3. It is essential that a competent staff of sufficient size is available to set up and operate the program.
4. A well-organized and planned predictive and preventive maintenance program should be tailored to individual plant needs. Program goals and a schedule for evaluating and reporting progress is desirable.
5. Written maintenance requests and orders are essential.
6. Predictive maintenance requires an adequate inventory of instrumentation for measuring parameters that indicate the current health of the operating facility.

DIAGNOSTIC TECHNIQUES

The use of diagnostic techniques should be practiced in order to minimize emergency breakdowns and increase the percentage of maintenance and repair that is scheduled. To evaluate the status of the current diagnostic techniques' effort, the following questions should be asked:

1. What percentage of the total maintenance effort is a result of breakdowns? Is this percentage greater than 15%? Has this percentage shown an increasing or decreasing trend in the past 3 years?
2. Is noise, vibration, and temperature instrumentation periodically utilized in order to determine the current status of important bearings throughout the plant?
3. Are highly stressed components of the various facilities throughout the plant periodically examined utilizing magnaflux or other instrumentation to determine if fatigue or stress failure is imminent?
4. Are important electrical circuits periodically tested for continuity, voltage drop, and so on, to assure their adherence to specifications?

APPLICATION OF THE COMPUTER

Today almost every segment of the maintenance control system can be improved through the sound use of the microprocessor. This facility should be utilized wherever it is cost-effective and where it can provide better reporting of useful information to management. In particular, the microprocessor will find extensive application in the development of standards, in scheduling, and in the development of regular management reports.

TRAINING

Training will improve the quality of the maintenance operations as well as the control operations and simultaneously increase the knowledge of those craftspersons who perform their responsibilities. Thus training will improve the performance of both hourly and salaried personnel. In order to have a measure of the need for a training program or to evaluate the quality of the existing training program, consideration should be given to the following:

1. Supervisors should be familiar with the concepts, principles, and procedures of the entire integrated maintenance management program.
2. Production personnel should be familiar with the purpose and functioning of the maintenance program.
3. Maintenance supervisors and craftspeople should know how to utilize the latest methods and techniques and understand the maintenance requirements of all newly acquired equipment.

COMPENSATION OF MAINTENANCE EMPLOYEES

A successful engineering maintenance management program is congruent with a satisfied work force, who are not only satisfied with their total compensation, but who are enthusiastic about their contribution to the welfare of the plant or business. The following questions should be asked periodically in order to evaluate the overall effectiveness of the wage payment plan:

1. Is the performance of the maintenance work force more than 100%?
2. Do a significant number of employees perform at more than 135%? (This would be an indication of loose standards.)
3. Is the labor turnover significantly high?
4. Of those employees who leave the company, are they the more skilled employees?
5. Are there a significant number of complaints related to compensation?

6. Do the total wages earned of each maintenance employee compare favorably with other industries in the area for the same class of work?
7. Are suggestions for methods improvement being received regularly from the maintenance work force?

REPORTS TO MANAGEMENT

It is necessary that reports to management be compiled in order to provide a continuing picture of the trend of plant maintenance costs and to show the position, progress, and effectiveness of the engineering maintenance management effort. Reports will assist management in making quick and reliable decisions that will sustain high maintenance efficiency. Some general principles that can be helpful in evaluating the quality of the reporting include:

1. Some reports should be presented in graphic form on a periodic basis so that the busy executive will be able to quickly see the trend over a period of time (see Fig. 17.1–17.3).

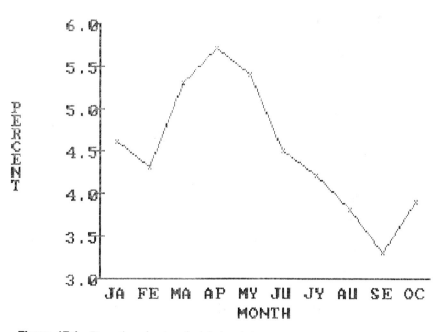

Figure 17.1 Downtime due to scheduled maintenance.

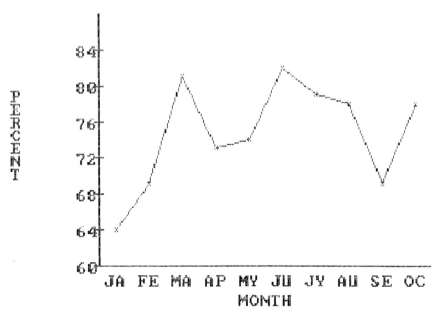

Figure 17.2 Percentage scheduled maintenance.

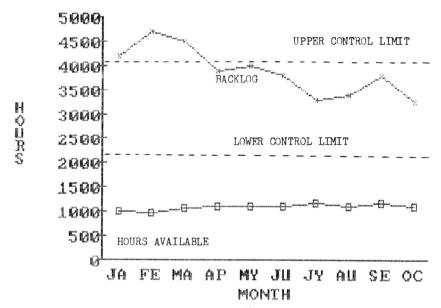

Figure 17.3 Backlog and hours available for maintenance work.

2. Important information such as machine downtime, backlog hours of scheduled maintenance work, and maintenance workmen efficiency should be reported regularly.
3. There should be adequate means established for accumulating the information necessary to produce and maintain important reports in a cost-effective manner.
4. There should be close liaison between the plant engineer and the plant controller in both the preparation and interpretation of the reports.

Maintenance Reports

Maintenance reports should be integrated. Reports developed for line supervision should be subsidiary to those reports designed for their immediate supervisors, the foremen. Similarly, the foremen's reports should be summarized and the pertinent information should flow into the next higher level of supervision.

Standard cost figures and budgets should be used to form basic benchmarks to which actual figures can be compared in the development of reports. Showing trends is important. Any figure standing by itself is seldom significant—it must be compared with something.

Budgets, usually are prepared annually, can be the basis for monthly cost reports that show

1. The total standard hours of service rendered
2. The actual hours of service provided
3. The ratio of standard hours of service rendered to the total standard production hours earned
4. The budget and actual expense for such costs as labor, materials and supplies, transportation, and fuel

It is recommended strongly that the computer be utilized in the development of reports. This tool permits the summarization and analysis of massive detail into meaningful and useful management reports that can highlight problem areas. For example, cost-effective action is invariably fruitful if repetitive maintenance work can be identified and then be subjected to operations analysis in order to develop ideal methods preparatory to establishing work standards based upon measurement. Other problems areas that need to be identified are those facilities where the maintenance cost is high when measured as a percent of replacement value and downtimes that impact on production.

Armstrong World Industries utilizes the following computer generated reports:

1. *PM Inspection Report.* Acts as a work order to initiate preventive maintenance inspections.

2. *PM Monthly Maintenance Summary.* Twelve month trend report by equipment expenditure.
3. *Open Job Report.* Weekly status report of all open jobs.
4. *Backlog Report.* Lists by craft the hours of backlog.
5. *Equipment Cost Record.* Lists all jobs performed to date on a piece of equipment with associated hours and dollars (labor, overhead, material, and total).
6. *Closed Job Report.* Monthly summary of jobs closed with complete breakdown of labor and material.
7. *Completed Job Status Report.* Monthly summary of jobs completed with associated hours and costs.
8. *Estimating Efficiency Report.* Monthly report comparing estimated hours to actual hours for each job.
9. *Stores Report.* Catalog of inventory items, brief description of items, unit prices, usage, current inventory.
10. *Monthly Maintenance Budget and Variance Report.*
11. *Monthly Maintenance Expense Report.*

The format of the Open Job Report, Monthly Maintenance Summary, Equipment Cost Record, Estimating Efficiency Report, Priority Report, and Completed Work Order Report are illustrated by Figs. 15.4 through 15.9.

Key Maintenance Indices

There are several key maintenance indices that need to be considered in the preparation of certain reports. These indices are assigned to one of three categories related to the general objectives of the engineering maintenance management program: maintenance administration, maintenance effectiveness, and maintenance cost.

Maintenance administration

1. Overtime hours per month.

$$\% = \frac{\text{Total overtime hours worked}}{\text{Total hours worked}} \times 100$$

2. Maintenance work orders planned and scheduled daily.

$$\% = \frac{\text{Work orders planned and scheduled}}{\text{Total work orders executed}} \times 100$$

(Text continues on page 352.)

Area XX

Operation XXX

Equipment Class XXX

Week Ending _____   Open Job Report   Plant XX

Job Number	Date Issued	Date Wanted	Priority	Description	Estimated Hours	Total Estimate	Actual Hours	Labor $	Material $	Outside $	Total
XXXXX	XXXX	XXXX	X	XXXXXX	XXXXXX	XXXX	XXXX	XXXX	XXXX	XXXX	XXXX

Notes:

1. List all jobs that are open.
2. Round off hours to nearest dollar.
3. Round off dollar figures to nearest dollar.
4. Issue report weekly.
5. Sort: (1) Area, (2) Operation, (3) Equip. Class, (4) Job Number.
6. Take job off of print-out when completed. Job remains open in EDP Files for 60 days.
7. Report printed on 14" paper.

"D's" & "M's"
Print out on line below job.

*(Indicates cost has reached or exceeds 80% of estimate)
**(Indicates cost has reached or exceeds 100% of estimate)

XXXX   Name or code of the engineer responsible for job.

Figure 17.4 Typical open-job report as issued regularly by a specific computer control system.

Monthly Maintenance Summary

Period XX-XX-XX

Plant X X

Area X X

Operation X X X

Equip. Class	Description	Amount Expended Last Year	Amount Expended Year-to-Date	Budget Year-to-Date	1	2	3	4	5	Monthly Costs 6	7	8	9	10	11	12
XXX	XXXXXXX	XXX	XXX	XXX	XXX	XXX		XXX		XXX	XXX	XXX	XXX	XXX	XXX	XX
Total Operation		XXX	XXX	XXX	XXX	XXX		XXX		XXX	XXX	XXX	XXX	XXX	XXX	XX

Notes:

1. Cost includes labor and materials.
2. Monthly cost columns show current month's cost in column 1 and the previous eleven month's cost in columns 2 to 12.
3. Round off figures to nearest dollar.
4. Report issued monthly.
5. Sort: (1) Area, (2) Operation, (3) Equipment Classification.
6. Report printed on 14" paper.

Figure 17.5 Computer-generated monthly report providing maintenance costs of specific equipment.

Equipment Cost Record

Operation XXX   Equip. Class XXX   Description XXXXXXXXXXXXX   Plant XX

Job Order	Description of Work	Class Wk.	Comp. Date	Hours	Labor	Dollars	Material Dollars	Total Cost
XXXXXXXX	XXXXXXXXXXXXXXXXX	XXX	XXXX	XXXX		XXXX	XXXXX	XXXX
	Equip. Class Totals			XXXX		XXXX	XXXXX	XXXX

Notes:
1. Round off hours to nearest tenth of hour.
2. Round off cost figures to nearest dollar.
3. Report issued monthly and is cumulative for current year.
4. Sort: (1) Operation, (2) Equipment Classification, (3) Work Order Number.
5. Report printed on 11" x 8 1/2" paper.

Figure 17.6 Computer-generated report providing cost information on the repair, maintenance, and modernization of capital equipment.

351

Estimating Efficiency Report　　　　　　　　Plant No.　X X
　　　　　　　　　　　　　　　　For Period Ending X X–X X–X X
Planner Code　X

Maintenance Area	Completed Work Order	Estimated Hours	Actual Hours	Estimated Efficiency
X	X X X X X X	X X X X	X X X X	X X X

Number Work Orders With Efficiency

Less than – 10 %	X X X
From – 10 to 10 %	X X X
Greater than 10 %	X X X

Notes

1. One sheet for each Planner.
2. Round off hours to nearest tenth.
3. Round off percentages to nearest whole percent.
4. Report issued monthly.
5. Sort (1) by Planner, (2) by Maint. Area,
   (3) by Work Order Number.
6. Formula for computing estimating efficiency:

$$\frac{\text{Estimated Hours} - \text{Actual Hours}}{\text{Estimated Hours}} \times 100 =$$

+ or – % Estimating Efficiency (+ over estimated; – under estimated)

7. Report printed on 8 1/2" x 11" paper.

Figure 17.7　Computer-generated maintenance labor efficiency report.

3. Scheduled hours versus hours worked as scheduled.

$$\% = \frac{\text{Hours worked as scheduled}}{\text{Total hours scheduled}} \times 100$$

4. Scheduled hours versus total hours worked.

$$\% = \frac{\text{Hours scheduled}}{\text{Total hours worked}} \times 100$$

<u>Priority Report</u>   <u>Plant No. X X</u>
                                               <u>For Period Ending</u> X

<u>Production Area</u> X X

	Hours Expended Current Month
Priority E	X X X X X
Priority 1	X X X X X
Priority 2	X X X X X
Priority 3	X X X X X
Priority 4	X X X X X
Total Hours	X X X X X

<u>Notes:</u>

1. One report for each Production Area.
2. Report issued monthly.
3. Round off figures to nearest hour.
4. Sort (1) by Production Area, (2) by Priority.
5. Report printed on 11 1/2" x 8" paper.

Figure 17.8 Computer-generated report providing information as to the number of maintenance work hours expended at various priority levels in each production area of the plant.

5. Craftsmen activity level.

$$\% = \frac{\text{Standard hours earned}}{\text{Total clock time}} \times 100$$

6. Current backlog (in crew-weeks).

$$\text{Crew weeks} = \frac{\text{Work scheduled ready to release (in man-hours)}}{\text{One crew-week (in man-hours)}}$$

7. Total backlog (in crew-weeks).

$$\text{Crew-weeks} = \frac{\text{Total man-hours of work awaiting execution}}{\text{One crew week (in man-hours)}}$$

Completed Work Order Report

Plant No. XX

Operation XXX   Equip. Class XXXXXXXXXXXXXXXXXXX   Date Issued XX-XX-XX   Date Comp. XX-XX-XX

Job Description XXXXXXXXXXXXXXXXXXXXXXX   Class Wk. XXX   Priority X   Area XX   Job No. XXXX

Breakdown of Costs

Labor

So.	Skill Code	Est. Hrs.	Act. Hrs.	Amount
12	1	XXXX	XXXX	XXXX
	16			
	Total Labor	XXXX	XXXX	XXXX

Materials

		Amount
So. 13	Stores Materials (List by Stores Number)	XXXX
So. 13	Purchased Materials (List by P.O. Number) Quantity XXXX	XXXX
	Total Materials	XXXX
So. 23	Outside Maintenance (List by P.O. Number)	XXXX
	Total Outside	XXXX
	Grand Total (Labor, Mat'ls, & Outside)	XXXX

1. One sheet, or sheets, for each completed job.
2. Round off hours to nearest tenth hour.
3. Round off cost figures to nearest dollar.
4. Report issued monthly.
5. Print report when job is completed. Job remains open in EDP files for 60 days. Issue new report at final closure if job accrues additional cost during this period.
6. Sort: (1) Operation, (2) Equip. Class, (3) Job Number.
7. Report printed on 8 1/2" x 5 1/2" paper.

Figure 17.9 Computer-generated completed maintenance work order report, giving details of cost of performing the work.

8. Predictive and preventive maintenance coverage.

$$\% = \frac{\text{Total man-hours of predictive and preventive maintenance}}{\text{Total man hours worked}} \times 100$$

Maintenance effectiveness

1. Emergency man-hours.

$$\% = \frac{\text{Man-hours spent on emergency jobs}}{\text{Total direct maintenance hours worked}} \times 100$$

2. Emergency and all other unscheduled man-hours.

$$\% = \frac{\text{Man-hours of emergency and unscheduled jobs}}{\text{Total maintenance man-hours worked)}} \times 100$$

3. Equipment downtime caused by breakdown.

$$\% = \frac{\text{Downtime caused by breakdown}}{\text{Total downtime}} \times 100$$

4. Evaluation of predictive and preventive maintenance.

$$\% = \frac{\text{Predictive and preventive inspections completed}}{\text{Predictive and preventive inspections scheduled}} \times 100$$

Also

$$\% = \frac{\text{Jobs resulting from inspections}}{\text{Inspections completed}} \times 100$$

Maintenance Costs

1. Maintenance cost per unit of production.

$$\text{Cost per unit} = \frac{\text{Total maintenance costs}}{\text{Total units produced}}$$

2. Inventory turnover rate per year.

$$\text{Rate} = \frac{\text{Annual consumption costs}}{\text{Average dollar inventory}}$$

3. Ratio of labor costs to material costs of maintenance.

$$\text{Ratio} = \frac{\text{Total maintenance labor costs}}{\text{Total maintenance materials costs}}$$

4. Percent supervision costs of total maintenance costs.

$$\% = \frac{\text{Total costs of supervision}}{\text{Total maintenance costs}} \times 100$$

5. Maintenance costs as percent of total manufacturing cost.

$$\% = \frac{\text{Total maintenance costs}}{\text{Total manufacturing costs)}} \times 100$$

6. Cost of a maintenance hour.

$$\$ = \frac{\text{Total cost of maintenance}}{\text{Total man-hours worked}}$$

7. Progress in cost reduction efforts.

$$\text{Index} = \frac{\% \text{ Maintenance man-hours spent on scheduled jobs}}{\text{Maintenance cost/Units of production}}$$

8. Labor cost to apply \$1.00 worth of materials.

$$\text{Labor cost} = \frac{\text{Labor dollars}}{\text{Materials dollars}}$$

9. Preventive maintenance costs as percent of total breakdown maintenance costs.

$$\% = \frac{\text{Total PM costs (including production losses)}}{\text{Total breakdown costs}}$$

Other periodic reports related to the efficiency of the PM program that may be desirable to prepare include

1. The effect of predictive and preventive maintenance expenditures on the total maintenance costs of selected items of equipment or selected areas in the plant
2. The impact of predictive and preventive maintenance expenditures on the amount of emergency and/or high-priority work on selected equipment

MAINTAINING GOOD COMMUNICATIONS

One of the principal reasons for weak performance of the overall engineering maintenance management program is poor communications. For example, many maintenance problems are designed into facilities, as was brought out in Chapter 11. These faults usually are the result of lack of communication between the design and maintenance groups. A breakdown in communication can result in improper maintenance with work performed either too late or unnecessarily early.

The various kinds of communication equipment that is available may give readers ideas as to how to improve communication in their particular plants or businesses. Of course, face-to-face meetings of plant engineers, maintenance engineers, maintenance supervisors, and line production supervisors provide the most effective communications. However, it is difficult to justify separate meetings for communication purposes on a regular basis involving these key people.

The principal communication equipment available today includes the following:

1. *Call bells.* A two-number code call is imposed on a wire circuit and sounds a code bell wherever the bells are installed. In addition, lights may be connected to the circuit in order to give a visual signal in addition to the audio signal.

2. *Sound-powered telephones.* These systems do not require the use of a battery or plug-in electric power. Stations are connected by wire, and power is provided by the spoken voice. (When the sound waves strike a transmitter diaphragm, they set up vibrations in a magnetic circuit.)

    These units are effective up to 30 miles and are used where hazardous conditions exist, such as in some rescue equipment and where power service is not reliable. The cost of these units has limited their use.

3. *Pneumatic tubes.* Here communication exists only between fixed points. This type of communication is effective for exchange of documents, laboratory samples, and so on.

4. *Amplified intercommunication systems.* These systems are economical and practical for both routine and emergency messages as long as someone is at or near the receiver. One of the systems can incorporate means to schedule a work load, assign personnel, and send back to a central console recordings of job starting and completion times. It is also able to alarm the console supervisor if a particular job is exceeding the estimated completion time.

5. *Written message relay system.* In this system, an electric circuit is used to transmit a written message or elementary sketch from one location to one or more other locations. The transmitter and receiver units are selfcontained and fully transistorized. Only an electric outlet and a pair of signal wires are needed.

6. *Public address.* The newer public address systems incorporate both public address and a reply system. Under this system, a number of selfcontained handsets are located throughout the plant together with supporting circuitry. The caller announces the intended receiver and this message is reproduced in all the other handsets and any auxiliary reproducers. The person called picks up the nearest handset and can call directly and confidentially to the caller.

7. *Radio paging.* Transistorized radio paging systems can pinpoint a key individual anywhere in the plant or even at some distance from the plant and give a complete private voice message. Low-frequency systems that do not require FCC licensing are recommended for an in-plant system. Each key person wears a microradio receiver and a clear alerting tone sounds when someone is calling the individual. Only the person called is alerted—no one else need be aware of the call.

   Very high-frequency systems are effective in buildings and for in-town calls. They require a central antenna and a FCC-licensed based station and are recommended for plants where buildings are widely separated and where there are outside loading operations.

8. *Two-way radio.* Two-way radio installations have been widely used on lift trucks, mine locomotives, tractors, and so on. Citizens band transistorized battery-operated walkie-talkie units are available at a nominal price. The source of power generally is a standard 9-volt battery.

9. *Closed-circuit television.* This facility is widely used to observe operations or instrumentation at remote locations from a centralized control console.

10. *In-transit electronic mail systems.* This system prints messages of up to 1000 alphanumeric characters on a roll of paper similar to ticker tape. The hardware can be programmed to beep or vibrate before it starts to print. The thermal print messages can be sent either to one or several units simultaneously. For example, centralized maintenance can communicate with several satellite maintenance centers at once. The communication distance is unlimited. Messages may be sent using any type of computer—from microcomputer to mainframe.

PARTICIPATION IMPROVEMENT GROUPS

Group participation, when properly administered, has had dramatic improvement results in a wide variety of businesses and industries. One of the most publicized applications of participating groups is the quality control circle which developed in Japan more than a decade ago. The opportunity to induce improvement in the maintenance function through the use of employee participation groups should not be overlooked.

A participation improvement group can be described as a small group of employees who meet regularly in order to identify and solve problems related to the company that they represent. Typically, a participation group would be made up of about ten employees. This would include the plant engineer, a representative from industrial engineering, the maintenance manager, and about seven maintenance technicians. Usually, all maintenance crafts are represented by the group of technicians. The maintenance manager or the plant engineer would chair the meetings which should not extend over two hours. The group normally would meet on a monthly basis and attendance would be required.

The industrial engineering representative would be utilized to assist in the collection of data and for the analysis of the facts obtained. With the attendance of the plant engineer, it would be possible to quickly develop solutions and provide implementation without the long delays often encountered when management approval is required. Thus, prompt action can usually be taken on all recommendations.

Since attendance is required, the entire group will participate regularly in all deliberations and assist in the decision making process that may impact on their work environment.

Records should be kept to verify the cost-effectiveness of the participation improvement group. Improvements can usually be confirmed through work sampling. This technique allows measurement of the reduction of waiting times, traveling times for tools and supplies, delays, and the increase in working times.

CONCLUSION

In order to improve the maintenance function continually, there should be a regular maintenance audit. This appraisal tool should be designed to measure what maintenance management does, how well it does it, and what it does not do that it should be doing. The 11 ingredients that comprise a sound engineering maintenance management program should be evaluated in the audit. Measures to determine the performance of each of these ingredients have been discussed. It should be understood that the maintenance management effort should

1. Measure its own action program
2. Determine the correlation between action and results
3. Monitor the effectiveness brought about by any changes introduced in each of the 11 ingredients
4. Maintain precise and clear communication between all facets of the maintenance, production, and management organizations

SELECTED BIBLIOGRAPHY

Nakajima, Seiichi, Yamashina, Hatime, Kumagai, Chitoku, and Toyota, Toshio. "Maintenance Management and Control." In Gavriel Salvendy, Ed., *Handbook of Industrial Engineering*, 2nd ed. New York: Wiley, 1992.

Tomlingson, P. D. "Evaluating Maintenance Performance." *Plant Engineering*, Vol. 42, No. 15 (1988).

Index

ABC classification system, 111–114
Abuse failures, 147–148
Access to maintenance, 242
Acquisition cost, 112
Allowances, time study, 70–72
Amplified inter-communication systems, 358
Anticipate breakdowns, 234–237
Anticipate life, 153–155
Applying allowances, 70–72
Apprenticeship agreement, 311
Apprenticeship training programs, 305–312
Audit, 360
Automatic lubricators, 173–175
Availability, 237–238

Backlog reports, 31–33
Bar coding, 122–123
Benchmark standards, 96–97

Bimetallic elements, 197
Boiler room preventive maintenance log, 187
Boiler systems, 262–264
Breakdowns, anticipating, 234–237
Brightness for categories of seeing, 260

Call bells, 358
Centralized maintenance, 6–12
Centralized storerooms, 115–116
Centrifuging, 172
Chance failures, 147–148
Charts
    control, 55–60
    ISO 9000 profile, 45
    oil selection viscosity, 166
    organization, 7, 9–11
    "p", 56–60
    PERT, 140

Classifying jobs, 325–327
Closed-circuit television, 304
Code card for machine tools, 18–19
Coding, bar, 121–123
Cogeneration, 272
Communication, maintaining,
    358–360
Communication systems
    amplified intercommunications
        systems, 358
    call bells, 358
    closed circuit television, 359
    in-transit electronic mail systems,
        359
    pneumatic tubes, 358
    public address, 359
    radio paging, 359
    sound-powered telephones, 358
    two-way radio, 359
    written message relay system,
        359
Compensation for maintenance
    work, 316–338, 344–345
Compensation plan, 333–337
Compound amount factor, 104
Compressors, installation of,
    279–282
Computerized maintenance
    cost accumulation and reporting,
        222, 344
    equipment and facilities control,
        205–209
    flexible manufacturing systems,
        209–221
    inventory control, 123–124,
        219–221
    lubrication schedule, 208
    performance reporting, 222–225
    reports to management, 222–225
    scheduling, 143–144

[Computerized maintenance]
    spare parts control, 219–221
    work control, 210–219
Conservation of energy, 253–273
Continuing education programs,
    313–314
Contractors
    cost appraisal, 284
    insurance, guarantees, and liens,
        288–289
    limitations, 283–284
    selecting, 285–286
    types of contracts, 286–288
    working with, 284–285
Control charts, 181
Control materials, 107–108
Control systems, 14–47, 340
Cost
    accumulation, 222
    acquisition, 112
    carrying, 112
    considerations, 2–3
    control systems, 340
    degradation, 3
    direct, 2
    importance of, 38–39
    indices, 356–357
    invested capital, 103
    lost production, 2
    materials in maintenance work,
        100–110
    possession, 112
    profile, 32
    reduction, 42–43
    standby, 3
    work order, 37–39
Costing of stock, 106–107
Costing of stores, 101–106
Craft performance report, 34
Critical path, 139–140

Curves
  adding, 79
  drilling, 79
  exponential, 81–82
  hyperbolic, 80
  learning, 28–31
  linear form, 80
  mortality, 148
  normal, 95
  parabolic, 80–81
  Pareto's law, 113
  plotting, 75–85
  polynomial, 82

Daily work scheduling, 128–130
Daily work time card, 26
Data, observing and recording,
    60–62
Decentralized mainenance, 6–12
Decentralized storerooms, 115–116
Degradation costs, 3
Design for maintainability
  design considerations, 230–242
  identification, 246
  interchangeability, 247–248
  objectives, 232
  principles, 238–242
  safety, 247–249
  specifications, 250
  standard practice procedures,
    242–250
  system effectiveness, 237–238
Diagnostic techniques
  dimension monitoring, 200–201
  electrical measurements, 193–194
  hydraulic testing, 196
  improvement of, 343
  laser beam, 200
  motion pattern monitoring,
    200–201

[Diagnostic techniques]
  nondestructive testing, 201–202
  optical alignment, 201
  pneumatic testing, 196
  process parameter monitoring,
    193
  shock-pulse measurement, 201
  sound intensity, 194–196
  stroboscopic motion analyzer,
    200
  temperature analysis, 196–199
  thermography, 196–199
  training for, 203–204
  tribology, 199–200
  ultrasonics, 195
  vibration analysis, 190–192
  videotape systems, 200
Dimension monitoring, 200–201
Direct costs, 2
Dispatch board, 141–143
Dispatching procedure, 20–22
Doors, 267–269
Downtime classifications, 185
Drilling time curves, 79
Drop forge preventive maintenance
    model, 177–182

Economic order quantity, 118–121
Effectiveness of system, 237–238
Efficiency reports, 31–35
Electrical layout, 277–279
Electrical measurements, 193–194
Electrical systems, 198
Electronic infrared sensors, 262
Emergency maintenance, 41
Energy conservation, 253–273
Energy recovery, 272
Energy utilization index, 256
Engineered maintenance
    management programs, 39–43

Equipment failure reports, 37–38
Equipment history, 37, 39–40,
    158–159
Equipment listing, 207
Estimating equipment reliability,
    234
Estimating maintenance work, 41,
    64–98
Estimating material costs, 100–110
Expected return, 103
Exponential curves, 81–82, 233

Failure, causes of in electronic
    equipment, 231
Failure reports, 38
Failure zone, 153–154
Fanning equation, 280
File, inventory, 115
Fire protection
    classification of fires, 292
    fighting procedure, 292
    organization, 291
    roles, 291
Fittings, lubrication, 175–176
Flexible manufacturing systems,
    209–210
Float time, 139
Flow of maintenance activity, 27
Fluorescent penetrant inspection,
    201
Footcandle requirements, 260
Forms
    analysis of job, 213
    coding sheet, 206
    planning, 127
    preventive maintenance check,
        161
    time study, 67–68
    work order, 21, 211–214
    work sampling, 57, 61

Foundations, concrete, 282
Frequencies of inspection, 162
Functional layouts, 276–277

Gamma distribution, 97–98
Gas thermometer, 197
General repairs, 125
Grease transfer, 168

Heating systems, 262–270
Heating zone map, 266
Heat loss
    boilers, 262–270
    doors, 267–269
    hot water systems, 263–264
    skylights, 267–269
    windows, 267–269
Heat transfer, 265–270
History of equipment, 37, 39–40,
    158–159
Hydraulic testing, 196
Hyperbolic curve, 80

Identification, for maintenance
    work, 246
Improvement groups, 360
Indices, 348–358
Induction training, 43
Infant mortality, 147–148
Information flow, 25
Infrared sensors, 198–199
In-plant training, 301–304
Inspection, 159–162
Inspection frequencies, 162
Inspection route, 162–163
Installation of compressors and
    pumps, 279–282
Instrument availability, 203
Insulation, 264–266

Integrated maintenance system, 5
Interchangeability, 246–247
In-transit electronic mail systems, 359
Inventory control, 108–109, 111–124, 219–221, 341–342
Inventory, cost of, 103
Inventory physical, 121–122
Inventory, record procedure, 114–115
Inventory, three categories of, 113–114
ISO 9000, 43–46

Job analysis, 316–320
Job assignment schedule, 139–141
Job card, 24
Job classification, 325–327
Job evaluation, 320–333
Job instructions, 186
Job rating, substantiating data, 323–324

Key indices, 348–358

Lamp performance data, 258–259
Laser beam, 200
Layout, plant types, 276–277
Lead times, 117–118
Learning curve, 28–31
Life of equipment, 153–155
Lighting, 256–262
Linear form of curves, 80
Logic monitors, 193–194
Log-log paper, plotting, 29
Long-range scheduling, 130
Lost production costs, 3
Lubricants, 176–177

Lubrication
    area travel times, 164
    automatic lubricators, 173–175
    fittings, 175–176
    lube oil analysis, 167
    methods of application, 167–175
    oil viscosity, 166
    schedule, 208
    standard practice procedures, 175
Luminous efficacy, 257

Machine records, 158–159
Magnetic particle inspection, 201–202
Maintainability, 232–246
Maintenance
    access to, 242
    audit, 360–361
    centralized, 6–12
    classifications, 125–126
    compensation for, 316–338, 344–345
    computerized, 209–221, 344
    computerized inventory control, 123–124
    control of materials, 107–108
    control systems, 14–47
    cost considerations, 2–3
    cost indices, 356–357
    cost profile, 32
    cost reduction, 42–43
    craftsmen performance report, 34
    decentralized, 6–12
    degradation costs, 3
    diagnostic techniques, 193–200, 343
    direct costs, 2
    dispatch board, 141–143
    dispatching procedure, 20–22
    effectiveness indices, 356

[Maintenance]
emergency, 41
engineered management
programs, 39–46
equipment records, 37, 39–40,
158–159
estimating, 64–98
flow of activity, 27
history of equipment, 37, 39–40,
158–159
improvement of performance,
339–360
information flow, 25
inventory control, 111–124,
341–342
job card, 24
lost production costs, 2
machine records, 158–159
material control procedures,
107–108
material costs, 100–110
measurement, 64–98, 341
measuring performance, 339–361
organization, 1–13
organization chart, 7, 9–11
participating improvement group,
360
place of, in organization, 6
planned, 39–41, 163–166
planning and scheduling,
125–145, 342
predictive, 15–16, 182–187, 343
preventive, 15–16, 146–187, 343
relation between quantity and
cost, 4
reliability-centered, 148–150
reports, 25–28
responsibility of, 3–6
skill development, 43
slotting standards, 93–98

[Maintenance]
standard data, 74
standby costs, 3
subcontracted services, 282–289
time reporting, 24–25
total productive, 151–152
training, 43, 299–315, 344
universal standards, 95–98
work order, 21
work order dispatching
procedure, 20–22
work order register, 23
work order system, 16–20
Manufacturing progress function,
28–31
Master scheduling, 128–130
Material control procedures,
107–108
Material costs, 100–110
Maynard Operation Sequence
Technique (MOST), 90–93
Mean forced outage time, 234–237
Mean time between failures,
234–237
Measurement of maintenance work,
64–98, 341
Mercury thermometer, 197
Metal reflectors, 271
Method determination, 3
Methods-Time Measurement
(MTM), 85–93
Misuse failures, 147–148
Mortality curve, 148, 234
MOST, 90–93
Motion pattern monitoring, 200–201
MTM-2, 85–90

National Electrical Code, 278–279
National Electrical Manufacturers
Association, 321

National Fire Protection
Association, 278
Nondestructive testing
fluorescent penetrant, 201
magnetic particle, 201–202
radiographic, 202
Nonstock items, 101–106
Normal curve, areas of, 95

Observing data, 60–62
Oil viscosity selection chart, 166
One-for-one incentive, 337
Optical alignment, 201
Optical pyrometer, 198
Order quantity, 118–121
Organization chart, 7, 9–11
Organization evaluation, 339–340
Overtime hour reporting, 222

Parabolic curve, 80–81
Pareto's law, 112–113
Participating improvement groups,
360
Performance reports, 31–35,
222–225
PERT, 138–141
Physical inventory, 121–122
Place of maintenance in plant
organization, 6
Planned maintenance, 39–41,
126–128
Planning form, 127
Plant availability profile, 33
Plant layout, 276–279
Plant rearrangement, 275–289
Plotting curves, 75–85
Pneumatic testing, 196
Pneumatic tubes, 358
Point job evaluation plan, 320–333

Polynomials, second-degree, 82
Possession costs, 112
Predictive maintenance, 15–16,
182–187
Pre-employment testing, 300–301
Preventive maintenance
characteristic of a plant in need
of, 155–158
check card, 187
improvement of, 343
inspection, 159–162
model, 177–182
objectives of, 155
questionnaire, 155–158
Process layouts, 276–277
Process parameter monitoring, 193
Product quality reports, 35–37
Program evaluation review
technique (PERT), 138–141
Programmable controllers, 194
Programmable lighting and
temperature selection, 261–262
Programmed instructions, 304–305
Public address communication, 359
Pumps, installation of, 279–282

Quality assurance management
system, 43–46
Quality circles, 360
Quality reports, 35–37
Quantity of maintenance and costs,
4
Questionnaire, for preventive
maintenance, 155–158

Radiation pyrometers, 198
Radiographic inspection, 202
Radio paging, 359
Record procedure, 114–115

Reduction of cost, 42–43
Regulation of residual waste, 294
Reliability, 232–234
Reliability-centered maintenance, 148–150
Reliability estimating, 234
Reliability improvement, 42
Reports
  backlog, 346
  cost of performance, 354–355
  downtime, 345
  efficiency, 35
  equipment failure, 37–38
  estimated versus actual hours, 224
  history, 37, 39–40
  management, 345–347
  monthly amount, 222–224
  overtime hours, 222
  percentage scheduled work, 346
  performance, 31–35
  plant availability, 222
  product quality, 35–37
Residual waste, controlling, 293–295
Responsibility of maintenance, 3–6
Robots, 189
Roofs, 269–270

Safety factor, 152–153
Safety precautions, designed into equipment, 247–249
Safety stock, 117–118
Salary job evaluation plan, 327–333
Scheduled maintenance, 125–145
Scheduling
  daily, 134–137
  improvement of, 342
  long range, 130
  PERT, 138–139

[Scheduling]
  requirements of, 137–138
  short interval, 137
  using the computer, 143–144
  weekly, 130–134
Screening, 3
Second-degree polynomials, 82
Seger cones, 198
Selection of stock, 106–107
Semilogarithmic paper, 83–84
Shock-pulse measurement, 201
Skill development, 43
Skylights, 267–269
Slotting standards, 93–98
Solar energy, 271–272
Sound intensity, 194–196
Sound-powered telephones, 358
Spare parts, 108–109, 219–221
Standard data, 73–75, 76–77
Standards, universal maintenance, 93–98
Standby costs, 3
Stock, 106–107
Stopwatch time study, 66–72
Storage methods, 116–117
Storerooms, 115–116
Stores, 101–106
Storm water discharge, 296–297
Straight-line layouts, 276–277
Stroboscopic motion analyzer, 200
Subcontracted services, 282–289
Synthetic lubricants, 176–177

Temperature measurement
  bimetallic elements, 197
  electrical systems, 198
  gas thermometer, 197
  infrared systems, 198–199
  mercury thermometer, 197
  optical pyrometer, 198

[Temperature measurement]
  radiation pyrometer, 198
  Seger cones, 198
  tempilsticks, 198
  thermistors, 197
  thermocouple, 197
  thermography, 198–199
  vapor-bulb thermometer, 197
Tempilsticks, 198
Thermistors, 197
Thermocouple, 197
Thermography, 198–199
Time card, 26
Time reporting, 24–25
Time study
  applying allowances, 70–72
  approach to operator, 69
  calculating the standard, 71–73
  dividing the job into elements,
    69–70
  equipment, 66
  form, 67–68
  observer's position, 69
  standard data, 73–75
  taking the study, 70
Time value of money, 103–104
Tolerance capability on new and
    worn facility, 36
Tool requirements for
    maintainability, 242–246
Training for diagnostics, 203–204
Training for engineering
    maintenance work
  apprentice agreement, 311
  apprentice training, 43, 305–312
  closed-circuit television, 304
  continuing education programs,
    313–314
  diagnostic training, 203–204
  in-plant training, 43, 301–304

[Training for engineering]
  pre-employment testing, 300–301
  programmed instruction, 304–305
  technician training, 303–304, 344
  technologist training, 303–304,
    344
  visual aid training equipment,
    303–304
  vocational technical schools,
    312–313
Triangular distribution, 178–182
Tribology, 190–200
Turret lathe standard data, 75
Two-bin inventory control, 117
Two-way radio, 359

U factor, 268–269
Ultrasonics, 195
Universal maintenance standards,
    95–98
Utilities management
  boiler/heating system, 262–270
  doors, 267–269
  energy utilization index, 256
  hot water systems, 263–264
  insulation, 264–266
  lighting, 256–262
  objectives, 254–256
  roofs, 269–270
  skylights, 267–269
  U factor, 268–269
  windows, 267

Vapor-bulb thermometer, 197
Vibration analysis, 190–192
Videotape for diagnostics, 200
Videotape facilities, 200
Visual aid equipment for training,
    303–304

Vocational technical schools, 312–313

Wearout, 147–148
Weekly scheduling, 130–134
Welding Institute standard data, 76–77
Welding time per inch of weld, 84
Windows, 267
Work ahead schedule, 128–134
Work control, computerized, 210–219
Work-Factor, 85
Work order, 21
Work order dispatching procedure, 20–22
Work order register, 23
Work order system, 16–20
Work request, 17

Work sampling
control charts, 56–60
definition, 48
designing the form, 56
determining the frequency of observations, 55–56
determining the observations needed, 54–55
form, 57
illustrative example, 50–53
information gathered by, 63
observing data, 60–62
planning the study, 53–54
recording the data, 60–62
theory of, 49
Work time card, 26

Zone, failure, 153–154